Quantum Teleportation Under Noise

Dongfen Li

Quantum Teleportation Under Noise

Key Technologies and Applications

Springer

Dongfen Li
Chengdu University of Technology
Chengdu, China

ISBN 978-3-031-91249-8 ISBN 978-3-031-91250-4 (eBook)
https://doi.org/10.1007/978-3-031-91250-4

This Springer imprint is published by the registered company Springer Nature Switzerland AG
The registered company address is: Gewerbestrasse 11, 6330 Cham, Switzerland

Preface

Information security is the fundamental prerequisite for the advancement of informatization, and cryptographic systems form the basis of information security. However, classical cryptographic algorithms, which rely on complex computations, are increasingly vulnerable to the advances in quantum computing and other powerful technologies in the post-Moore era. These classical algorithms can be easily compromised or rendered ineffective.

Quantum teleportation, based on the principles of indivisibility, the no-cloning theorem, and the instantaneous collapse of quantum states upon measurement, leverages theories of quantum coherence and entanglement. It enables the measurement of quantum states, the calculation of particle collapse states, and the reconstruction of unknown quantum states through unitary transformations. This ensures absolute security in information transmission and plays a critical role in large-scale quantum computing, long-distance quantum communication, and quantum networks. As such, quantum teleportation has become a research focus across academic, military, and information technology fields worldwide.

Given the increasing challenges posed by quantum decoherence under noise conditions, and the non-classical, non-local correlations of photons with multiple degrees of freedom, the traditional assumption of independent single degrees of freedom in quantum teleportation protocols is called into question. This undermines the efficiency of existing quantum teleportation channels, especially when limited resources and entanglement sudden death present constraints. To address these issues, this work proposes a robust quantum teleportation protocol that is better suited to practical environments. The main research achievements include:

Construction of a High-Fidelity Entangled Quantum Teleportation Channel

First, to address the issues of "entanglement sudden death" and quantum decoherence during quantum teleportation, the evolution of entanglement over time between two spatially independent atomic systems is analyzed. The relationship between the initial entanglement of the atoms and cavity field is explored to establish an immune model against entanglement death. Additionally, by analyzing the oscillatory phenomenon of the decoherence factor in the T-C and J-C models, a quantum decoherence entanglement evolution model is constructed. Furthermore, considering different noise characteristics, an immune noise model based on density matrix and a joint noise-immune model based on decoherence-free subspace (DFS) are designed, along with an analysis of optimal qubit efficiency for different roles. Finally, to tackle the issue of low channel capacity caused by entanglement death, channel capacity encoding schemes based on graph states, including tree and forest structures, are proposed. These schemes analyze coherence information and channel capacity, calculate the approximate noise capacity and noise tolerance, and identify regions where quantum information can be transmitted under various noise channels and shared local noise conditions.

Quantum Information Splitting of Bell State and Arbitrary State in Different Quantum Channels

To address the issues of transmitting a small number of particles and computational complexity in the same or different channels, an innovative quantum information splitting scheme for Bell states and arbitrary states using different quantum channels is proposed. First, a four-particle entangled state is used as a quantum channel to split two-particle Bell states, enabling the transmission of Bell state information. Second, quantum information splitting of arbitrary single-particle and two-particle states is achieved using a five-particle entangled state. Finally, different quantum channels, such as four-particle cluster states combined with two-particle Bell states or GHZ states, are used to split arbitrary three-particle states. The splitting process employs different measurement operations to achieve high-efficiency quantum information separation, with the physical implementation of the process demonstrated in cavity quantum electrodynamics (QED).

Multiple Degrees of Freedom Quantum Teleportation Protocol for Immune Noise

To address the high cost and low efficiency of single-degree-of-freedom quantum teleportation, a new controllable multi-degree-of-freedom quantum teleportation protocol immune to noise is proposed. First, the demand for multiple degrees of freedom is analyzed by guiding spin-orbit angular momentum to adjust their intensity distribution over time, space, and frequency. This enables the preparation of high-brightness entangled sources with multiple degrees of freedom. Second, the conversion relationship between spin-orbit angular momentum and hyper-Bell states is described, with hyper-Bell state measurements and superdense coding performed to meet multiple target requirements under angular momentum constraints. Finally, a controllable quantum teleportation identity authentication model is designed, achieving robust quantum teleportation under both independent and shared noise environments. The fidelity and average fidelity efficiency of the protocol are analyzed to enhance measurement efficiency and security.

Application of Fault-Tolerant Quantum Teleportation Protocol in the Noise Channel

To address practical bottlenecks in the application of quantum teleportation, two typical applications—noise-immune quantum dialogue and continuous-variable quantum key distribution—are designed. First, a dialogue mechanism based on decoy states and steganography is proposed to verify the identities of the communicating parties and the security of the communication channel. Single photons and generalized entangled states are constructed, with pre-shared identifiers converted into logical quantum states immune to joint noise. These states are randomly interwoven within the information sequence for both identity authentication and eavesdropping detection. Second, a deterministic continuous-variable quantum key distribution protocol is proposed, where the sender transmits predetermined keys to the receiver via a public channel. After verifying the security and performance of these applications, both are shown to exhibit high robustness and efficiency, with significant potential and fault tolerance under noisy conditions.

In summary, this research addresses the key challenges in quantum teleportation under noise conditions, enriching quantum communication theory and paving the way for practical applications of quantum teleportation technology.

Chengdu, China Dongfen Li

Contents

Chapter 1
Introduction

With the rapid development of Internet information, the Internet has gone deep into every corner of society. However, the problem of information leakage exposed in recent years makes information security face a huge crisis. Under the attack of malware and Trojan horse, the cracking of traditional classical encryption algorithms becomes easier, while the advent of quantum computing technology intensifies the "vulnerability" of classical encryption methods. Quantum communication uses the characteristics of indivisibility of quantum, non-cloning of state, and detection of instantaneous collapse. Based on the basic theories of quantum coherent superposition and quantum entanglement, it measures the quantum state and calculates the collapse state of particles and reconstructs the unknown quantum state information through unitary transformation, so as to ensure the absolute security of information transmission. Quantum teleportation plays an important role in large-scale quantum computing, long-distance quantum communication, and quantum computing network with its advantages of strong security, large storage capacity, and resource-saving. It has become a research hotspot in academic circles, military organizations, and information technology industries all over the world.

1.1 Research Background and Significance

With the vigorous development of mobile payment, e-commerce, Internet finance, and other emerging businesses, the tentacles of the Internet have penetrated into every corner of social and economic life. However, in recent years, the constant exposure of monitoring and eavesdropping scandals and user privacy leaks have further exacerbated the concerns of the public and all sectors of society about network and information security [1]. Information security incidents are increasing year by year, such as the "prism gate" incident, the "Jingdong Mall" hacking incident, and other incidents that reveal a large number of sensitive information and private data, which further shows the crisis faced by China's national cyberspace

D. Li, *Quantum Teleportation Under Noise*,
https://doi.org/10.1007/978-3-031-91250-4_1

security [2]. The central leading group for network security and informatization was established, and the first-class discipline of cyberspace security was set up, which promoted network and information security to a national strategic position. Therefore, whether from the perspective of individual needs or national strategies, the problem of network and information security has reached a moment that must be urgently solved.

However, the existing classical communication encryption methods are not as secure as expected. Their security is guaranteed by the computational complexity of cryptographic algorithms [3]. By constructing a huge amount of computation, they can convert between "plaintext" and "ciphertext" to protect the integrity of information. For example, the security of the RSA algorithm is reduced to the factorization of large integers, which is an NP-hard problem. According to Moore's law, the computing power of a processor will increase exponentially. The emergence of new technologies, such as distributed computing, cloud computing, and GPU, makes the computing speed fast. Under the attack of Trojans and hackers, it becomes easier to crack the classical encryption algorithm. The emergence of quantum computing technology exacerbates the insecurity of classical encryption. Quantum communication is based on the quantum uncertainty principle, the quantum non-cloning theorem, and the quantum mechanics principle of quantum entanglement. The quantum cryptography theory, which aims to provide unconditional secure (absolutely secure) information exchange, has attracted more and more attention. Its emergence has subverted the security of classical algorithms and completed a series of tasks that are difficult for classical cryptographic algorithms to achieve. For example, the quantum Shor's algorithm [4] can change the NP problem of large factor decomposition into a P problem. Quantum computing technology, with its huge information-carrying capacity and efficient parallel processing ability, is a powerful "spear" for cracking passwords [5].

As new technology in the post-Moore era [6], quantum communication is a particularly important research branch of quantum informatics. Based on the basic principles of quantum mechanics, it can achieve safe and reliable information processing and transmission that cannot be achieved by classical communication technologies. Theoretically, it can achieve absolutely secure communication. Because of its high transmission efficiency and absolute security, it has become the ultimate weapon to ensure network information security [7]. It will greatly change the development of traditional informatics, enrich the research methods of cryptography, promote research in information security-related fields such as national defense and finance, and provide new ideas and strong security guarantee for the research of confidential communication and informatization construction and is regarded as the leading science to ensure the future informatization [8]. From 2012 to 2014, the Nobel Prize in Physics for 3 consecutive years was awarded to scientists in quantum measurement and operation, Higgs boson discovery, and blue light-emitting diode invention [9], which are related to the quantum field, and quantum communication has shown signs of revolutionary breakthrough. It uses the basic characteristics of quantum coherent superposition and quantum entanglement to realize information transmission. The quantum state is non-clonable and indivisible, and the detection of instantaneous collapse ensures the absolute security of the transmission content, attracting more and more attention [10–12]. Therefore, quantum communication solves the two difficult problems that classical encryption algorithms are easy to crack and easy to monitor and is a powerful "shield" to ensure network and information security [13].

The development of quantum communication has been related to the security of national cyberspace, so most countries attach great importance to it and have invested a lot of human and material resources in research, which has been applied on different occasions. The first quantum communication key distribution network was built and put into use by the Defense Advanced Research Projects Agency in 2004 [14]. In the same year, Europe also launched research on "secure communication based on quantum cryptography" networks [15]. China has also made a series of achievements in quantum cryptography. Academician Guo's team built the world's first "quantum government network" in 2009 and put it into use [16]. In 2012, academician Pan's team built the "financial information quantum communication verification network" and the "Hefei metropolitan quantum communication experiment demonstration network" in Hefei [17, 18]. On August 16, 2016, the first quantum science experiment satellite "Mozi" independently developed by China was successfully launched in Jiuquan Satellite Launch Base [19], and the 4-month on-orbit test was completed on January 19, 2017, and officially delivered to the user unit for use [20]. On March 13, 2017, academician Pan revealed at the "two sessions" that the "Beijing-Shanghai trunk line," the world's first 1000-kilometer quantum communication backbone network, has been completed [21]. The outline of China's 13th 5-year plan has included quantum communication and quantum computer in one of the 15 major scientific and technological innovation projects, which shows that China attaches importance to quantum technology.

Quantum communication is divided into quantum key distribution (QKD) and quantum teleportation (QT) [22, 23]. QT is used to transmit quantum information. If there is a quantum channel between the communication parties, quantum information is transmitted on the quantum channel by quantum teleportation [24]. QT uses quantum entanglement distribution and quantum joint measurement technology to transmit an unknown quantum state to a distant place to realize the spatial transfer of quantum states without transmitting the physical carrier itself. It has the advantages of high reliability, low communication complexity, and resource-saving. The core resource of QT is quantum entanglement, that is, when the superposition principle of quantum states is embodied in a multiparticle or multiple degrees of freedom system, a strange phenomenon of quantum entanglement will occur [25–29]. When the physical system is coupled with the noise environment, it will accelerate the growth of quantum decoherence. At the same time, quantum entanglement, which is strongly dependent on coherence, will also appear in the process of entanglement decay, or even the sudden death of entanglement [30–35]. As a basic element of quantum information processing and an important element of practical quantum technology, QT is a very safe and reliable carrier for the transmission of secret information. It plays a vital role in large-scale quantum computing, quantum entanglement, long-distance quantum communication, and quantum computing network.

It has become a new research direction in the scientific field and has important research significance.

At present, the problems in the study of quantum teleportation are as follows:

1. Without a unified immune noise framework for high-fidelity entangled quantum teleportation channel, it is impossible to describe the entanglement evolution characteristics in the local common mode, and it is difficult to describe the entanglement evolution model and the cause of the sudden death of entanglement.
2. The existing channel capacity does not reach the theoretical value, and the channel utilization is low.
3. The measurement method is based on a single degree of freedom, and can only transmit a single degree of freedom quantum state, which is at the level of quantum basic theory. Because photons have multiple degrees of freedom characteristics such as wavelength, momentum, spin, and orbital angular momentum, and show the non-localized and nonclassical strong correlation in quantum physical systems, this shakes the assumption of a single degree of freedom independence in quantum teleportation theory [30, 36].

In a word, the existing research cannot optimize the three indicators of availability, channel utilization, and security of quantum teleportation at the same time, which makes it difficult for the existing quantum channel framework and channel capacity to provide satisfactory services to the system under the contradiction of limited resources, entanglement death, and decoherence, reduce the fidelity and efficiency of transmitted quantum states, and hinder the practicality of quantum teleportation [33, 37, 38]. As a necessary way to develop quantum computing and quantum network, practical quantum teleportation technology plays an important role in the fields of finance, government affairs, national defense and military, and long-distance communication (such as space exploration). At the same time, it will drive the rapid development of components and other upper and lower industrial chains and has become a research hotspot in the basic research field of quantum communication in recent years [39, 40].

The research on quantum teleportation protocol is a hot topic at present, but noise and multiple degrees of freedom bring great challenges to the existing quantum entangled teleportation channel framework and transmission support capability. This book seeks a new unified framework for entangled quantum teleportation channels with immune noise, explores the capacity coding theory of quantum teleportation channels based on immune noise channels, designs quantum information splitting schemes for the Bell states and arbitrary states in different channels, constructs a multi-free metric teleportation protocol, and obtains real observations on quantum dialogue and quantum key distribution of continuous variables. It has urgent practical significance and basic academic value.

1.2 Dynamic Analysis of Related Research

This section briefly describes the work related to the contents of this book at home and abroad and the key issues that need to be further studied from the four aspects of quantum entanglement evolution under noisy channels, quantum channel capacity and coding, and quantum information splitting and quantum teleportation protocol.

1.2.1 Correlation Between Quantum Entanglement Evolution and Immune Noise Model

The early research on the process of quantum teleportation was carried out in an ideal environment without paying attention to the impact of noise on quantum teleportation. It was not until Di Vincenzo [41] and Bennett [42], respectively, studied the feasibility of quantum computing, they found that noise was the main factor affecting the transmission of information. Subsequently, scholars described information under different quantum noises but were only limited to using error-correcting coding [43] and finding quantum non-decoherence-free subspace [34] to offset the impact of noise on the system. In 2004, when Yu et al. studied the entanglement evolution of the double-qubit entangled system in the process of noise decoherence, they found that the nonlocal entanglement decay process can be faster than the local decoherence decay process and the quantum entanglement can die out in a limited time, that is, the entanglement suddenly dies [44]. Soon, Almeida et al. verified the existence of this phenomenon through experiments [45]. After that, the researchers used the super operator summation master equation method to study quantum dynamics for different quantum noises and different systems. Carvahlo et al. characterized the entanglement evolution of three quantum GHZ states and W states by using the quantum master equation under the local independent heat reservoir and dephasing noise and analyzed their robustness [46].

In 2010, Siomau et al. used the same quantum state method as Carvahlo to describe the system evolution density matrix for thc local independent Pauli noise environment and found that the GHZ state and W state have different robustness in different Pauli environments [47]. In 2012, under the local independent multilateral noise, Siomau et al. described the system evolution density matrix of the three-qubit GHZ state and the noise cancellation model [48]. In 2016, Li et al. studied the influence of joint rotation noise and joint dephasing noise on physical quantum states by constructing a fidelity quantum teleportation model of immune collective noise. At the same time, taking cluster states as quantum carriers, they studied the logical quantum states immune to joint rotation noise and joint dephasing noise, respectively. The decoherence independent free subspace of the immune collective noise is constructed, so that the quantum state is still in the maximum entangled state after changing, so as to realize the fidelity of quantum teleportation that can resist the collective noise. The unitary evolution of the system under the joint rotation noise and joint dephasing noise is characterized [34]. At the same time, scholars have studied the influence of bit reversal noise and depolarization channel noise on the entanglement evolution of physical systems [49, 50].

These studies are based on the single degree of freedom and local noise environment, ignoring the characteristics of entanglement evolution, noise and entanglement measure under the influence of multiple degrees of freedom, and several common noises, and the immune model can only solve the local noise problem in a single isolated system and can only probabilistically reduce the decoherence and entanglement death of quantum states caused by noise. Because the quantum teleportation process needs to consume a lot of quantum resources, which inevitably reduces the channel capacity and cannot meet the needs of long-distance robust quantum teleportation, we improve the limitations of the existing research by constructing a unified framework of high-fidelity entangled quantum teleportation channels under these related characteristics.

1.2.2 Research on the Correlation Between Quantum Channel Capacity and Coding

In 1997, Schumacher et al. put forward the theorem based on quantum channel [51], mainly focusing on the quantum capacity and privacy capacity, which is developed from information theory. The study of quantum capacity is helpful to understand the capability and efficiency of quantum error-correcting codes in quantum processing. The study of privacy capacity is helpful to better understand the security and efficiency of quantum key distribution. In 2003, Devetak et al. gave the normal expression of privacy capacity and the defined value of "single-letter" non-flat lower bound [52]. The theorems of quantum capacity have been proposed [53]. Shor uses information theory to prove that the additivity conjecture of entanglement and the super additivity of entanglement formation are equivalent to the additivity conjecture of minimum input entropy of quantum channel [54]. Subsequently, in 2008, Winter et al. successively proposed the quantum capacity and privacy capacity of symmetric quantum channels and gave corresponding theorems and expressions [55]. Shadman et al. [56] use the advantages of asymmetric channels to solve the problem of decoherence in quantum channels, which makes it difficult to maintain the maximum entangled state and consumes entanglement information in this process. In 2013, Kesting calculated the noise tolerance of different quantum codes in the Pauli environment and obtained the optimal coding method of different quantum concatenated codes in the depolarized channel [57]. In the same year, Chen et al. proved the message authentication capacity under strong security conditions by using the information theory security and quantum random coding theory and the strong security capacity coding model based on the noise channel [24]. From 2014 to 2015, Li et al. exploited the properties of super entangled states that feature easy preparation, convenient measurement, and straightforward implementation of dense coding to construct a quantum covert channel for entanglement switching. They also proposed an efficient dense coding method based on super entangled switching, which effectively enhanced the channel utilization and capacity in quantum communication [33, 38–40]. Later, scholars popularized the Calderbank-Shor-Steane (CSS) construction theorem, which is simple and feasible, and analyzed the quantum channel capacity with entanglement assistance, which has been widely used [58, 59].

These research results are all single operations in time, space, or frequency. In the process of measurement and coding, auxiliary particles are introduced and unitary transformation matrix is constructed. The receiver cannot completely distinguish the transmitted quantum states, and the channel capacity has not reached the theoretical value, which cannot meet the needs of long-distance quantum teleportation.

1.2.3 Research on Quantum Information Separation

The concept of quantum information transmission was proposed by Bennet et al. in 1993. According to the principle of quantum mechanics, after the communication parties share the entangled pair, they can know the transmission of the unknown quantum state on another particle. Because of its unconditionally safe and timely transmission of quantum states, it has developed rapidly in theory and experiment, and it has become a very popular research point in the field of quantum communication. In 2001, H. J. Briegel and R. Raussendorf proposed a new quantum entangled state cluster state. This new cluster state requires that the number of entangled particles in the quantum channel be greater than three. Cluster states have some special properties. The persistence of entanglement is better than that of states. Cluster states contain the properties of class [60, 61] and class [62–65] entangled states. Because cluster states have the largest correlation and entanglement persistence, its entanglement is more difficult to be destroyed, and cluster states have been widely used in quantum teleportation [66, 67], dense coding [68, 69], quantum computing [70–73], and quantum information separation [74, 75]. Among these methods, quantum information splitting is a common method to realize quantum information processing, because it can transfer information between three and more parties through entangled states and classical communication channels. Since Hillery first proposed quantum information separation using states, quantum information separation has attracted people's attention and has been widely used in experiments [76–80]. Quantum states include Bell states [81], cluster states (four-particle, five-particle, six-particle), and the GHZ states, which can be considered as quantum channels. After that, Zhan et al. [82] proposed quantum teleportation based on high-dimensional entangled states. Yin et al. [83] proposed quantum teleportation based on four-particle cluster states. Nie et al. [84] proposed a scheme for quantum information splitting through GHZ state as a quantum information splitting scheme for separating any single particle through a quantum channel [74, 84–89]. In 2014, Li et al. [90] proposed a quantum information splitting scheme using four-particle cluster states and GHZ states to split any three-particle states. In the same year, Li et al. proposed a quantum information splitting scheme using seven-particle entangled states to split arbitrary three particles and a quantum information separation scheme using four-particle entangled states to split two-particle Bell states [37, 91].

1.2.4 Research on Quantum Teleportation Protocol

From 1994 to 1996, Davidovich and Bennett proposed a new scheme of quantum teleportation based on the joint measurement of Bell states [92, 93]. In 1997, Austrian scientists Bouwmeester and others successfully realized entanglement-based quantum teleportation for the first time [94]. In 1998, Rome of Italy and Kimble of the United States used the continuous-variable theory to conduct quantum teleportation of coherent light field and nuclear magnetic resonance [95, 96], which was listed as one of the top ten scientific and technological advances in the United States that year.

In 2002, Sangchul Oh et al. proposed a fidelity quantum teleportation scheme in a local independent quantum noise environment, analyzed the average fidelity and safety efficiency, and opened a precedent for the study of quantum teleportation on noise channels [97]. In 2003, Bowen et al. of the Australian National University successfully experimented [98]; Takei et al. realized quantum teleportation in the vacuum state of different field modes based on the properties of squeezed states [99]. In 2008, Jung et al. used the three-particle state or W state as the quantum channel, proposed the single-qubit quantum teleportation theory in the depolarization environment and the local independent Pauli environment, and creatively found that the selection of the quantum channel depends on the noise environment. It is concluded that under local independent Pauli noise, GHZ state and W state have different and more suitable quantum channels with different selected parameters. However, under local independent depolarization noise, the same transmission effect can be obtained by selecting the GHZ state and W state as quantum channels [100].

In 2010, Pan Jianwei and others from the University of Science and Technology of China conducted another quantum teleportation experiment in a free space of 16 km, and the experimental results were successfully published on the cover of *Nature Photonics* [101]. From 2011 to 2013, Hu and other scholars used four-particle cluster state, two-particle Bell state, three-particle GHZ state, and W state as quantum channels to analyze their fidelity under local independent high temperature, zero temperature, and dephasing noise [102, 103]. Since 2015, Li and other scholars have used cluster states, GHZ states, and other particle states to conduct quantum teleportation in local independent joint dephasing noise, joint "rotation" noise, bit reversal channel, depolarization channel, amplitude damping channel, channel, and other environments and analyzed different entanglement and other indicators [33–40, 104–106]. For continuous-variable quantum information, Seshadreesan et al. proposed a scheme for teleportation of quantum states in non-Gaussian entangled states and Schrodinger cat states [104]. At the same time,

Pan Jianwei and others broke through the limitations of a single degree of freedom and creatively realized the quantum teleportation experiment under multiple degrees of freedom. The experimental results were published in nature, providing a strong experimental guarantee for promoting the research on quantum transmission under multiple degrees of freedom [6].

In 2016, Zuppardo et al. put forward and verified the over-distribution theory of quantum entanglement in a noisy environment. The over-distribution of quantum entanglement may be the only way to realize the entanglement revenue [107]; Xiao et al. proposed a scheme of enhanced quantum teleportation by adjusting different measured parameters in the damped noise channel. The decoherence effect can be eliminated through the combination of partial measurement and reversal after local measurement [108]. Based on the quantum superposition and entanglement principle, Li et al. carried out a small quantum state transmission experiment of frozen bacteria and microorganisms by using nuclear magnetic resonance equipment to further apply quantum teleportation [108].

Combined with the entanglement source preparation and Bell state measurement characteristics in noise channel, there have been many studies on the theory and experiment of quantum teleportation, but these theories are based on the independent assumption of single degree of freedom, which cannot meet the requirements of long-distance robust quantum teleportation. Therefore, the characteristics of spin-orbit angular momentum with multiple degrees of freedom and the new method of super Bell state measurement resources must be introduced into the quantum teleportation theory to solve the problem in the new situation.

1.3 Challenging Scientific Issues

From the analysis of relevant research trends at home and abroad in Sect. 1.2, it can be seen that the existing research on quantum teleportation, and the entanglement evolution of quantum teleportation under noise channels, has caused four key scientific problems, as follows:

1. Characterization of decoherence and entanglement death of quantum states in quantum channels.

There will be quantum "entanglement sudden death" and quantum state decoherence in the quantum teleportation channel, which are not only the basis of modeling and analysis of the immune noise model but also the basis of realizing multiple degrees of freedom quantum teleportation protocol and quantum information separation and also the basis of studying a new framework of quantum teleportation channel based on high-fidelity entanglement. Therefore, we will first establish the quantum entanglement evolution model under the noise channel; analyze the "sudden death of entanglement" problem, the quantum decoherence problem, and the entanglement under local common noise in the process of quantum communication; and construct the "death of entanglement" model, the quantum decoherence model, and the quantum entanglement evolution model under the local common noise. This work requires not only the analysis and research foundation of quantum entanglement dynamics but also strong data abstraction ability. Therefore, how to characterize the decoherence and "entanglement death" of quantum states caused by noise in the entangled quantum teleportation channel is a key problem to be solved. For this problem, the third chapter describes the evolution of quantum entanglement under "entanglement death," quantum decoherence, and local common mode and constructs a framework of high-fidelity entangled quantum teleportation channel, which effectively solves this problem.

2. Channel capacity theory considering multiple degrees of freedom demand intensity in noisy channel.

Measuring the parameter index of quantum teleportation in terms of channel performance and quality of service under noisy channels is the premise to promote the basic theory of quantum teleportation and solve the problem of network information security. At the same time, we also provide a theoretical basis for the optimization scheme design of the multiple degrees of freedom quantum teleportation protocol. Based on the study of the immune noise model, we will introduce the idea of quantum graph state cascade coding, combine the security of information theory and quantum random coding method, and consider the characteristics of multiple degrees of freedom and in-depth research on quantum channel capacity coding. The difficulty of this book is to further encode and popularize the tree graph and forest graph through the graph state base. Therefore, how to find a new method of quantum graph state-dense coding based on an immune noise channel is the second scientific problem to be solved. For this problem, the third chapter proposes a graph-based channel capacity coding method, which solves the problems of channel performance and quality of service.

3. Solving quantum information separation based on Bell state and arbitrary state in different channels.

Based on the framework of the quantum communication channel, according to the different characteristics of particles, the quantum information separation scheme based on the Bell state and arbitrary state is designed. By using entangled state and cluster state as quantum channels to split quantum information of arbitrary particles and Bell state, the quantum information splitting process with high efficiency, high security, and low cost for different numbers of particles through different measurement methods is the core problem of quantum teleportation. Selecting different quantum channels (such as entangled state, cluster state, W state, etc.) for information splitting of a single particle, two particles, and three particles is the key research point. At the same time, the quantum information splitting of different types of particles in different channels provides the basis for the application of quantum key distribution with continuous variables, which has important research and application value. Therefore, how to characterize the efficient quantum information separation of different types of particles in different channels is the third scientific problem to be solved. For this problem, the fourth chapter splits Bell states and quantum information based on arbitrary states (arbitrary single particle, arbitrary two particles, and arbitrary three particles) in different channels.

4. Solving transmission mechanism based on multiple degrees of freedom in quantum teleportation.

Based on the unified framework of high-fidelity entangled quantum teleportation channel, according to the characteristics of photons in multiple degrees of freedom, a new multiple degrees of freedom quantum teleportation transmission mechanism is designed. By adjusting the distribution of photons in time, space, and frequency, the multiple requirements of photons under spin-orbit angular momentum to meet the preparation of high-brightness entangled sources and the measurement of hyperentangled Bell states are achieved, which are the key problems to be solved in quantum teleportation. Quantum teleportation under spin-orbit angular momentum freedom is the core feature of this study. At the same time, quantum teleportation protocol provides a basis for further design of quantum dialogue and continuous-variable quantum key distribution optimization applications and has important practical application value. Therefore, how to characterize quantum entanglement and super Bell state measurement in spin-orbit angular momentum multiple degrees of freedom is the third scientific problem to be solved. For this problem, the controllable multiple degrees of freedom quantum teleportation protocol with immune noise is designed in Chap. 5. In Chap. 6, the fault-tolerant quantum teleportation application under the noise channel is designed to obtain the real observation of quantum dialogue and quantum key distribution of continuous variables.

1.4 Research Content and Innovation

Clarifying research ideas lays a good foundation for research content, determining research content to support the completion of research objectives, and refining innovation points to improve research height.

1.4.1 Research Route

A unified framework model, coding method, and overall solution are proposed by combining theoretical modeling, software simulation, and physical system verification. By making full use of the existing research results of quantum information splitting and security dialogue, modeling analysis and optimization design, as well as cryptography and information system security, and absorbing the frontier research ideas and technologies in related fields, combined with the fault-tolerant quantum teleportation business data and user needs in noise channels, the research methods of quantum teleportation are combined with traditional cryptography, and typical applications are selected as specific research objects to carry out the basic theoretical research and key technology design in this book. Specific research ideas are as follows:

1. Research on quantum teleportation characteristics analysis and theoretical modeling based on secure information private communication.

Based on the requirements of secure information private communication, aiming at the four scientific problems of quantum teleportation under noise channel, we study the channel framework of quantum teleportation with high-fidelity entanglement, the quantum information splitting of Bell states and arbitrary states in different channels, the quantum teleportation channel capacity of immune noise model, and the fault-tolerant quantum teleportation application under noise channel; explore the ways to enhance the fidelity, capacity, and security of quantum teleportation; and provide theoretical support for reducing noise interference and improving the fidelity and security of quantum information.

2. Introducing a multidisciplinary research perspective to study entanglement evolution and channel capacity.

Based on the entanglement evolution and information capacity of quantum teleportation under noise channel, a new high-fidelity entangled quantum teleportation channel framework and coding method are established by introducing an open perspective of multidisciplinary research and using the analysis methods in information theory coding and cryptography.

3. Evaluation and validation of theoretical models and key methods for secure applications in noisy channels.

The theoretical work is evaluated and verified by the combination of software simulation and actual system experiments. Firstly, the model is preliminarily verified and corrected by simulation, and furthermore, convincing verification and evaluation are carried out through experiments in real environments to ensure the credibility of theoretical work.

1.4.2 Research Content

Noise and multiple degrees of freedom have brought great challenges to the basic theory and application technology of quantum teleportation and also brought opportunities for basic innovation in theory and application. We focus on source innovation and examine the basic problems in quantum teleportation, such as the construction of the immune model, the preparation of a high-brightness entanglement source, the measurement of the super Bell state, and the relationship between channel capacity requirements from the new perspective of the entanglement evolution of the fusion of noise and multiple degrees of freedom. We study the remote dense quantum teleportation mechanism of multiple degrees of freedom under the noise channel, establish a new unified high-fidelity quantum teleportation channel framework, design efficient and safe quantum state cascade coding, construct a new quantum teleportation protocol, and explore the new development of quantum information theory under the noise channel.

The key problems of quantum teleportation are examined from the new perspective of "entanglement death" in noise channels and the multiple degrees of freedom characteristics of the quantum system. Based on the main line of "safe quantum channel framework construction → quantum information splitting → channel capacity coding → quantum teleportation protocol establishment → typical application design," quantum teleportation protocol is designed to explore the new development of quantum teleportation.

The quantum channel framework, quantum channel capacity, and quantum teleportation are the core of this book, and the contents of each part are gradually progressive from bottom to top. The preparation of the entanglement source and the experimental platform of multiple degrees of freedom quantum teleportation are indispensable tools in this book, which provide corresponding means to verify and improve the four scientific problems in the research process. The channel framework of entangled quantum teleportation provides the basis for the application of teleportation. These two parts provide a design basis for quantum teleportation, and further research on teleportation based on channel framework and channel capacity can be carried out. At the same time, different research contents can also promote each other. For example, through the actual system deployment guidance mechanism, more sufficient observation of the relationship between quantum channel and channel capacity and invisible transmission can be obtained to guide the correction and improvement of the theoretical model. Specifically, we carry out the following four aspects of research, and the research content and relationship are shown in Fig. 1.1. The specific research contents are described as follows:

1. High-fidelity entangled quantum teleportation channel framework.

How to understand the introduction of high-fidelity entangled quantum teleportation channel under the noise channel is the basic problem. Firstly, the relationship between the "sudden death of entanglement" problem and quantum decoherence is analyzed, and thc immune quantum teleportation channel framework is further constructed. Based on the entanglement characteristics produced by the coupling between the external environment and the quantum state of the system, the sudden death of entanglement and quantum decoherence are studied. The influence of different noises on the entanglement evolution of entangled mixed states in the local common mode is analyzed, and the entanglement evolution model of the density matrix of the open quantum system is constructed. Based on quantum entanglement evolution and unitary transformation, the noise characteristics of the real physical system coupled with the environment are analyzed, and the coherent state of multiple degrees of freedom under noise is studied. The framework of high fidelity and high security of immune noise is designed, and the dynamic switching model of immune different noise is constructed. Because the existing noise channel coding scheme is to complete simple quantum communication under a single degree

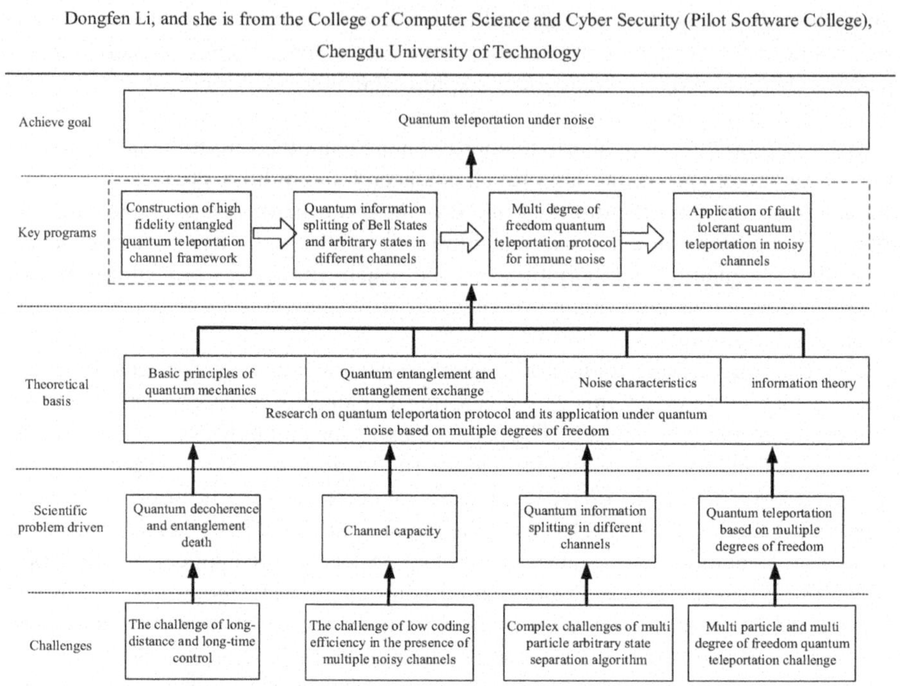

Fig. 1.1 The content and relationship of the research

of freedom, the transmission efficiency is low, the channel capacity cannot meet the requirements of remote transmission, and the cost is high. The quantum cascade coding method is used to effectively calculate the approximation value of the quantum capacity of the noise channel, the calculation speed, and the noise tolerance of the quantum information of the channel state transmission. Since its calculation speed is three orders of magnitude higher than that of the Monte Carlo algorithm, the utilization rate of the channel capacity reaches the highest value.

2. Quantum information splitting between Bell states and arbitrary states in different channels.

Based on the high-fidelity entangled quantum teleportation channel framework, according to the different characteristics of particles, a quantum information splitting scheme based on Bell state and arbitrary state is designed. By selecting different quantum channels (such as entangled state, cluster state, W state, etc.), the information splitting of a single particle, two particles, and three particles is carried out. For different numbers of particles, different measurement methods are used to achieve high efficiency, high security, and a low-cost quantum information splitting process. Firstly, four-particle entangled states are used as quantum channels to split quantum information of two-particle Bell states. Secondly, the five-particle entangled state is used to split the quantum information of arbitrary single-particle state and two-particle state and four-particle cluster states and the two-particle Bell state to split arbitrary three-particle state. Finally, the process of quantum information separation is physically realized in cavity quantum electrodynamics.

3. Quantum teleportation protocol based on multiple degrees of freedom.

Based on the immune noise model and high-capacity quantum channel, taking into account the multiple degrees of freedom characteristics of quantum wavelength, momentum, spin angular momentum, and orbital angular momentum, the controllable identity authentication model is designed through the preparation of high-brightness entanglement source and Bell state measurement under the degree of freedom of spin-orbit angular momentum. Considering the multiple degrees of freedom characteristics of the quantum physics system, the preparation of high-brightness quantum entanglement source under multiple degrees of freedom is realized by the characteristics of quantum state under spin angular momentum and orbital angular momentum. Based on the research on the preparation of high-brightness entanglement source with multiple degrees of freedom, the progressive relationship between the preparation of entanglement source and the Bell state measurement is described. A variety of measurement methods under multiple degrees of freedom are studied. The super entangled Bell state base measurement, enhanced local measurement, and dense measurement models are designed to realize safe quantum teleportation under multiple degrees of freedom.

4. Fault-tolerant quantum teleportation application in the noise channel.

The existing quantum teleportation is only suitable for an ideal environment or single quantum transmission, and the measurement technology is based on a single degree of freedom. For example, the quantum information is split by a single degree of freedom, and the receiver reconstructs the original information according to the set measurement basis through certain measurements, unitary transformation, and density matrix between the communicators. Based on the above research, this book can further design two application schemes, including quantum dialogue and quantum key distribution of continuous variables.

1.4.3 Innovation

Our contributions and innovations are summarized as follows:

1. A unified high-fidelity entangled quantum teleportation channel framework is constructed for the first time, and under this framework, a quantum teleportation protocol with entanglement evolution, high efficiency, robustness, and immune noise is designed. On this basis, the quantum dialogue of immune noise and quantum key distribution of continuous variables are realized, which provide a new idea for the development of quantum teleportation from theory to practical application in noise channels.

2. The idea of concatenated code is used to construct secure quantum state coding with high utilization and multiple degrees of freedom, which can effectively calculate the approximation value of quantum capacity of noise channel, the calculation speed, and the noise tolerance of quantum information transmitted through channels.
3. The controllable multiple degrees of freedom quantum teleportation protocols such as time and space in noise channels are innovatively proposed. The distribution characteristics of multiple degrees of freedom such as time and space are analyzed. The multiple degrees of freedom demand intensity is adjusted by guiding the spin-orbit angular momentum, and the high-brightness entanglement source is effectively prepared, and the super Bell state measurement is carried out to ensure the efficiency of multiple degrees of freedom.
4. The splitting of Bell state and arbitrary state quantum information on different channels is innovatively constructed, which provides new ideas for quantum teleportation applications such as quantum dialogue and quantum key distribution.
5. Quantum teleportation is realized by quantum dialogue, and continuous-variable quantum key distribution is realized, which opens up a new application field for further promoting practical quantum teleportation technology.

1.5 The Organizational Structure of Content

This book is divided into seven chapters. The first chapter is the introduction, the second chapter is the theoretical basis, the third to sixth chapters are the core research content, and the seventh chapter is the summary and outlook. The chapter structure is arranged as follows:

The first chapter is the introduction. By analyzing the irreplaceable role of quantum communication in the field of network and information security, we expound on the background and significance of this study. We analyze the relevant research trends and put forward some challenging scientific problems. Finally, it introduces the innovation and the organizational structure of the chapter arrangement.

The second chapter is the theoretical basis. Firstly, three basic principles commonly used in quantum teleportation are introduced, including the state superposition principle, the uncertainty principle, and the non-cloning theorem. Secondly, the basic concepts of quantum entanglement and entanglement swapping, entangled states, and related knowledge of entanglement swapping are introduced. Finally, the quantum noise characteristics of open systems are introduced, including the evolution process of open systems, measurement methods, and the measurement of quantum fidelity. In short, this chapter lays a theoretical foundation for the relevant schemes in Chaps. 3 to 6.

In Chap. 3, the framework of high-fidelity entangled quantum teleportation channel is constructed. The quantum entanglement evolution model, immune noise model, and channel capacity coding under noise are constructed.

In Chap. 4, the quantum information separation of Bell states and arbitrary states in different channels is studied. Using four-particle entangled states to split two-qubit Bell states and using a variety of different entangled channels to split arbitrary particle states to split quantum information.

In Chap. 5, a multiple degrees of freedom quantum teleportation protocol for immune noise is designed: Prepare, measure, and code the high-brightness entanglement source, authenticate the controllable quantum teleportation, describe the quantum teleportation process, and analyze the efficiency of the protocol.

In Chap. 6, the application of quantum teleportation is realized, that is, quantum dialogue and quantum key distribution of continuous variables.

Chapter 7 contains the summary and outlook. This chapter makes a brief summary and analysis of the research and expounds on the shortcomings of the existing work and the problems to be solved. At the same time, the future work is prospected.

References

1. C.C. Huang, K.J. Farn, A study on E-Taiwan Promotion Information Security Governance Programs with E-government Implementation of Information Security Management Standardization. Int. J. Netw. Secur. **18**(3), 565–578 (2016)
2. N.S. Safa, R.V. Solms, S. Furnell, Information security policy compliance model in organizations. Comput. Secur. **56**(1), 70–82 (2016)
3. S. Pirandola, J. Eisert, C. Weedbrook, A. Furusawa, S.L. Braunstein, Advances in quantum teleportation. Nat. Photon. **9**(10), 5439–5439 (2015)
4. L.K. By, Quantum Mechanics helps in searching for a needle in a haystack. Phys. Rev. Lett. **79**(2), 325–328 (1997)
5. A.K. Lenstra, H.W. Lenstra, M.S. Manasse, et al., The number field sieve, in *Proceedings of the Twenty-Second Annual ACM Symposium on Theory of Computing*, (ACM, New York, 1990), pp. 564–572
6. P.W. Shor, *Algorithms for Quantum Computation: Discrete Logarithms and Factoring* (IEEE Computer Society Press, New Mexico, 1994), pp. 124–134
7. H. Takesue, Quantum communication experiments over optical fiber, in *Principles and Methods of Quantum Information Technologies*, (Springer, Tokyo, 2016), pp. 53–70
8. Y. Yu, H.A.A. Man, G. Ateniese, et al., Identity-based remote data integrity checking with perfect data privacy preserving for cloud storage. IEEE Trans. Inf. Forensics Secur. **99**, 1–13 (2017)
9. http://baike.sogou.com/v89396.htm
10. C.E. Shannon, 保密系统的通信理论 (Communication Theory of Secrecy System). 贝尔实验室技术杂志 (Bell Syst. Tech. J.), **28**(4):656–715, 1949
11. H.K. Lo, H.F. Chau, Unconditional security of quantum key distribution over arbitrarily long distances. Science **283**(5410), 2050–2056 (1999)
12. P.W. Shor, J. Preskill, Simple proof of security of the BB84 quantum key distribution protocol. Phys. Rev. Lett. **85**(2), 441–444 (2000)
13. D. Gottesman, H.K. Lo, N. Lutkenhaus, et al., Security of quantum key distribution with imperfect devices. Quantum. Inf. Comput. **4**(5), 325–360 (2004)

14. C. Elliott, A. Colvin, D. Pearson, et al., Current status of the DARPA quantum network, in *Proceedings of SPIE, Orlando, FL, USA*, vol. 5815, (2005), pp. 138–149
15. A. Poppe, M. Peev, O. Maurhart, Outline of the SECOQC quantum-key-distribution network in Vienna. Int. J. Quantum. Inf. **6**(2), 209–218 (2008)
16. 许方星, 陈巍, 王双, 等 (F. Xu, W. Chen, S. Wang et al.), 多层级量子密码城域网 (Multi-tiered Quantum Cryptography Metropolitan Area Network). 科学通报 (Chin. Sci. Bull.) **54**(16), 2277–2283 (2009)
17. 王蔚蔚 (W. Wang), 金融信息量子通信验证网开通 (Financial Information Quantum Communication Verification Network Launched). 合肥日报, 2012年2月23日 (Hefei Daily, February 23, 2012)
18. 宗合,宗媛 (H. Zong, Y. Zong), 合肥城域量子通信试验示范网全网开通仪式举行 (Full Network Launch Ceremony Held for Hefei Metropolitan Quantum Communication Trial Demonstration Network). 合肥日报, 2012年3月31日 (Hefei Daily, March 31, 2012)
19. 刘诗瑶 (S. Liu), 我国成功发射世界首颗量子卫星 (China Successfully Launches World's First Quantum Satellite). 人民日报, 2016年8月16日 (People's Daily, August 16, 2016)
20. 喻思娈 (S. Yu), 首颗量子卫星"墨子号"在轨交付 (First Quantum Satellite 'Micius' Delivered for Orbital Operation). 人民日报, 2017年1月19日 (People's Daily, January 19, 2017)
21. 黎静, 谷薇薇 (J. Li, W. Gu), 量子通信"京沪干线"将开通 (Quantum Communication 'Beijing-Shanghai Backbone' to be Operational). 合肥晚报, 2017年3月13日 (Hefei Evening News, March 13, 2017)
22. X.L. Wang, Quantum teleportation of multiple degrees of freedom of a single photon. Nature **518**(7540), 516–519 (2015)
23. H. Ryszard, H. Paweł, H. Michał, H. Karol, Quantum entanglement. Rev. Mod. Phys. **81**(1), 865–942 (2009)
24. Z. Margherita, K. Tanjung, P. Tomasz, B. Somshubhro, B. Anindita, D. Prasenjit, H. Saronath, M. Kavan, P. Mauro, Excessive distribution of quantum entanglement. Phys. Rev. **93**(1), 012305 (2016)
25. L. Banchi, S.L. Braunstein, S. Pirandola, Quantum Fidelity for Arbitrary Gaussian States. Phys. Rev. Lett. **115**(26), 260501 (2015)
26. P.C. Obando, F.M. Paula, M.S. Sarandy, Trace-distance correlations for X, states and the emergence of the pointer basis in Markovian and non-Markovian regimes. Phys. Rev. A **92**(3), 676–677 (2015)
27. V. Špička, T.M. Nieuwenhuizen, P.D. Keefe, Physics at the FQMT'11 conference. Phys. Scr. **151**(1), 014001 (2012)
28. K. Inoue, Quantum noise in parametric amplification under phase-mismatched conditions. Opt. Commun. **366**(1), 71–76 (2016)
29. E. Oelker, T. Isogai, J. Miller, M. Tse, L. Barsotti, N. Mavalvala, M. Evans, Audio-band frequency-dependent squeezing for gravitational-wave detectors. Phys. Rev. Lett. **116**(4), 041102 (2016)
30. M. Elamraoui, G. Gadret, J.C. Jules, J. Fatome, C. Fortier, F. Désévédavy, I. Skripatchev, Y. Messaddeq, J. Troles, L. Brilland, W. Gao, T. Suzuki, Y. Ohishi, F. Smektala, Microstructured chalcogenide optical fibers from As2S3 glass: towards new IR broadband sources. Opt. Express **18**(25), 26655–26665 (2010)
31. G.F. Shi, Bidirectional quantum secure communication scheme based on Bell states and auxiliary particles. Opt. Commun. **283**(24), 5275–5278 (2010)
32. S. Chatrchyan, V. Khachatryan, A.M. Sirunyan, et al., Observation and studies of jet quenching in PbPb collisions at sNN=2.76 TeV. Phys. Rev. C **84**(2), 1859–1864 (2011)
33. D.F. Li, R.J. Wang, F.L. Zhang, Z.G. Qin, E. Baagyere, Splitting unknown Qubit State using Five-Qubit Entangled State. Int. J. Theor. Phys. **55**(4), 1962–1972 (2016)
34. R.J. Wang, D.F. Li, F.L. Zhang, Z.G. Qin, E. Baaguere, H.Y. Zhan, Quantum dialogue based on hypertanglement against collective noise. Int. J. Theor. Phys. **55**(8), 1–9 (2016)

35. D.F. Li, R.J. Wang, F.L. Zhang, E. Baagyere, Quantum information splitting of arbitrary two-qubit state by using four-qubit cluster state and Bell-state. Quantum Inf. Process. **14**(3), 1103–1116 (2015)
36. D.F. Li, R.J. Wang, F.L. Zhang, Quantum information splitting of a two-qubit Bell state using a four-qubit entangled state. Chin. Phys. C **39**(4), 043103 (2015)
37. D.F. Li, R.J. Wang, F.L. Zhang, Quantum information splitting of arbitrary three-qubit state by using four-qubit cluster state and GHZ-state. Int. J. Theor. Phys. **54**(4), 1142–1153 (2015)
38. D.F. Li, R.J. Wang, F.L. Zhang, F.H. Deng, Quantum information splitting of arbitrary three-qubit state by using seven-qubit entangled state. Int. J. Theor. Phys. **54**(6), 2068–2075 (2015)
39. R.J. Wang, D.F. Li, Z.G. Qin, An Immune Quantum Communication Model for dephasing noise using Four-Qubit Cluster State. Int. J. Theor. Phys. **55**(1), 609–616 (2015)
40. R.J. Wang, D.F. Li, F.L. Zhang, Z.G. Qin, E. Baaguere, H.Y. Zhan, Quantum dialogue based on hypertanglement against collective noise. Int. J. Theor. Phys. **55**(8), 3607–3615 (2016)
41. D.P. Di Vincenzo, Quantum computation. Science **270**(5234), 255–261 (1995)
42. C.H. Bennett, Quantum information and computation. Phys. Today **48**, 24–30 (1995)
43. D.A. Lidar, D. Bacon, K.B. Whaley, Concatenating decoherence-free subspaces with quantum error correcting codes. Phys. Rev. Lett. **82**(22), 4556–4559 (1999)
44. T. Yu, J.H. Beverly, Finite-time disentanglement via spontaneous emission. Phys. Rev. Lett. **93**(14), 140404 (2004)
45. M.P. Almeida, F. Melo, M. HorMeyll, A. Salles, S.P. Walborn, P.H. SoutoRibeiro, L. Davidovich, Environment-induced sudden death of entanglement. Science **316**(5824), 579–582 (2007)
46. G. Adesso, A. Serafini, F. Illuminati, Multipartite entanglement in three-mode Gaussian states of continuous-variable systems: quantification, sharing structure, and decoherence. Phys. Rev. A **73**(3), 032345 (2006)
47. M. Siomau, S. Fritzsche, Entanglement dynamics of three-qubit states in noisy channels. Eur. Phys. J. D Atom. Mol. Opt. Plasma Phys. **60**(2), 397–403 (2010)
48. M. Siomau, Entanglement dynamics of three-qubit states in local many-sided noisy channels. J. Phys. B Atom. Mol. Opt. Phys. **45**(3), 035501 (2012)
49. F. Caruso, V. Giovannetti, C. Lupo, S. Mancini, Quantum channels and memory effects. Rev. Mod. Phys. **86**(4), 1203 (2014)
50. A. Mazhar, Robustness of genuine tripartite entanglement under collective dephasing. Chin. Phys. Lett. **32**(6), 060302 (2015)
51. B. Schumacher, M.D. Westmoreland, Sending classical information via noisy quantum channels. Phys. Rev. A **56**(1), 131–138 (1997)
52. I. Devetak, The private classical capacity and quantum capacity of a quantum channel. IEEE Trans. Inf. Theory **51**(1), 44–55 (2005)
53. G. Smith, Private classical capacity with a symmetric side channel and its application to quantum cryptography. Phys. Rev. A **78**(2), 022306 (2008)
54. P.W. Shor, Equivalence of additivity questions in quantum information theory. Commun. Math. Phys. **246**(3), 453–472 (2004)
55. G. Smith, J.A. Smolin, A. Winter, The quantum capacity with symmetric side channels. IEEE Trans. Inf. Theory **54**(9), 4208–4217 (2008)
56. Z. Shadman, H. Kampermann, C. Macchiavello, D. Bruß, Optimal super dense coding over noisy quantum channels. New J. Phys. **12**(7), 073042 (2010)
57. A. Winter, Tight uniform continuity bounds for quantum entropies: conditional entropy, relative entropy distance and energy constraints. Commun. Math. Phys. **347**(1), 291–313 (2016)
58. F. Kesting, F. Fröwis, W. Dür, Effective noise channels for encoded quantum systems. Phys. Rev. A **88**(4), 042305 (2013)
59. A. D'Arrigo, G. Benenti, G. Falci, C. Macchiavello, Information transmission over an amplitude damping channel with an arbitrary degree of memory. Phys. Rev. A **92**(6), 062342 (2015)

60. Y.Y. Nie, Y.H. Li, A.S. Wang, Semi-quantum information splitting using GHZ-type states. Quantum Inf. Process. **12**(1), 437–448 (2013)
61. 叶天语,蒋丽珍 (T. Ye, L. Jiang), 可控量子秘密共享协议窃听检测虚警概率分析 (Analysis of false alarm probability in eavesdropping detection for controlled quantum secret sharing protocol). 光子学报 (Acta Photon. Sin.) **41**(9), 1113–1117 (2012)
62. Y.B. Sheng, L. Zhou, S.M. Zhao, Efficient two-step entanglement concentration for arbitrary W states. Phys. Lett. A **85**(4), 042302 (2012)
63. 李渊华, 刘俊昌, 聂义友 (Y. Li, J. Liu, Y. Nie), 基于W态的跨中心量子网络身份认证方案 (W-state based cross-center quantum network identity authentication scheme). 光子学报 (Acta Photon. Sin.) **39**(9), 1616–1620 (2010)
64. 周小清, 邬云文 (X. Zhou, Y. Wu), 三光子纠缠W态隐形传输令牌总线网的保真度计算 (Fidelity Calculation for W-state Teleportation Token Bus Network with Three-photon Entanglement).光子学报 (Acta Photon. Sin.) **39**(11), 2093–2096 (2010)
65. D. Ping, X. Zheng-yuan, Y. Ming, et al., Generation of cluster states. Phys. Rev. A **73**, 33818 (2006)
66. L. Ye, G.C. Guo, Probabilistic teleportation of an unknown atomic state Chin. Phys. **11**, 996–998 (2002)
67. Y. Ming, Probabilistic teleportation of a four-particle cluster state. J. Yanbian Univ. (Nat. Sci.) **35**(2), 137–140 (2009)
68. N. Gisin, G. Ribordy, W. Tittel, et al., Quantum cryptography. Rev. Mod. Phys. **74**(1), 145 (2002)
69. L. Ye, G.C. Guo, Scheme for implementing quantum dense coding in cavity QED. Phys. Rev. A **71**(3), 034304 (2005)
70. S. Muralidharan, J. Kim, N. Ltkenhaus, et al., Ultrafast and fault-tolerant quantum communication across long distances. Phys. Rev. Lett. **112**(25), 250501 (2014)
71. J.R. Samal, M. Gupta, P.K. Panigrahi, et al., Non-destructive discrimination of Bell states by NMR using a single ancillaqubit. J. Phys. BAt. Mol. Opt. Phys. **43**, 095508 (2010)
72. D.D.B. Rao, S. Ghosh, P.K. Panigrahi, Generation of entangled channels for perfect teleprotation using multielectron quantum dots. Phys. Rev. A **78**(4), 042328 (2008)
73. E.S. Prasath, S. Muralidharan, C. Mitra, et al., Multipartite entangled magnon states as quantum communication channels. Quantum Inf. Process. **11**(2), 397–410 (2012)
74. Y.Y. Nie, Y.H. Li, J.C. Liu, et al., Quantum information splitting of an arbitrary three-qubit state by using two four-qubit cluster states. Quantum Inf. Process. **10**(3), 297–305 (2011)
75. S. Choudhury, S. Muralidharan, P.K. Panigrahi, Quantum teleportation and state sharing using a genuinely entangled six-qubit state. J. Phys. A **42**(11), 115303 (2009)
76. Z.X. Man, Y.J. Xia, N.B. An, Genuine multiqubit entanglement and controlled teleportation. Phys. Rev. A **75**(5), 052306 (2007)
77. S. Muralidharan, P.K. Panigrahi, Perfect teleportation, quantum-state sharing, and superdense coding through a genuinely entangled five-qubit state. Phys. Rev. A **77**(3), 032321 (2008)
78. F.G. Deng, X.H. Li, C.Y. Li, P. Zhou, H.Y. Zhou, Multiparty quantum-state sharing of an arbitrary two-particle state with Einstein–Podolsky–Rosen pairs. Phys. Rev. A **72**(4), 044301 (2005)
79. K. Hou, Y.B. Li, S.H. Shi, Quantum state sharing with a genuinely entangled five-qubit state and Bell-state measurements. Opt. Commun. **283**(9), 1961–1965 (2010)
80. S. Muralidharan, P.K. Panigrahi, Quantum information splitting using multipartite cluster states. Phys. Rev. A **78**(6), 062333 (2008)
81. D.F. Li, R.J. Wang, F.L. Zhang, Quantum information splitting of a Two-Qubit Bell state using a Four-Qubit Entangled state. Chin. Phys. C **39**, 010601 (2014)
82. Z. Zhao, Y.A. Chen, A.N. Zhang, T. Yang, H.J. Briegel, J.W. Pan, Experimental demonstration of five-photon entanglement and open-destination quantum teleportation. Nature **430**, 54 (2004)
83. X.F. Yin, Y.M. Liu, W. Zhang, Z.J. Zhang, Simplified four-qubit cluster state for splitting arbitrary single-qubit information. Commun. Theor. Phys. **53**, 49–53 (2010)

84. Y.Y. Nie, Y.H. Li, J.C. Liu, M.H. Sang, Quantum information splitting of an arbitrary three-qubit state by using two four-qubit cluster states. Quantum Inf. Process. **10**(3), 297–305 (2011)
85. Y.Y. Nie, Y.H. Li, J.C. Liu, M.H. Sang, Quantum information splitting of an arbitrary three-qubit state by using a genuinely entangled five-qubit state and a Bell-state. Quantum Inf. Process. **11**(2), 563–569 (2012)
86. Y. Nie, Y. Li, J. Liu, et al., Quantum state sharing of an arbitrary four-qubit GHZ-type state by using a four-qubit cluster state. Quantum Inf. Process. **10**, 603–608 (2011)
87. Y.Y. Nie, Y.H. Li, J.C. Liu, M.H. Sang, Opt. Commun. **284**(5), 1457–1460 (2011)
88. Y. Nie, Y. Li, J. Liu, et al., Quantum information splitting of an arbitrary three-qubit state by using a genuinely entangled five-qubit state and a Bell-state. Quantum Inf. Process. **11**(2), 563–569 (2012)
89. Y.Y. Nie, Z.H. Hong, Y.B. Huang, X.J. Yi, S.S. Li, Theor. Phys. **48**, 1485 (2009)
90. D.F. Li, R.J. Wang, F.L. Zhang, F.H. Deng, E. Baagyere, Quantum information splitting of arbitrary two-qubit state by using four-qubit cluster state and Bell-state. Quantum Inf. Process. **14**(3), 1103–1116 (2015)
91. L. Davidovich, N. Zagury, M. Brune, J.M. Raimond, S. Haroche, Teleportation of an atomic state between two cavities using nonlocal microwave fields. Phys. Rev. A **50**(2), R895 (1994)
92. C.H. Bennett, G. Brassard, S. Popescu, B. Schumacher, J.A. Smolin, W.K. Wootters, Purification of noisy entanglement and faithful teleportation via noisy channels. Phys. Rev. Lett. **76**(5), 722–725 (1996)
93. D. Bouwmeester, J.W. Pan, K. Mattle, M. Eibl, H. Weinfurter, A. Zeilinger, Experimental quantum teleportation. Nature **390**(6660), 575–579 (1997)
94. S.L. Braunstein, H.J. Kimble, Teleportation of continuous quantum variables. Phys. Rev. Lett. **80**(4), 869–972 (1998)
95. A. Furusawa, J.L. Sørensen, S.L. Braunstein, C.A. Fuchs, H.J. Kimble, E.S. Polzik, Unconditional quantum teleportation. Science **282**(5389), 706–709 (1998)
96. S. Oh, S. Lee, H. Lee, Fidelity of quantum teleportation through noisy channels. Phys. Rev. A **66**(2), 022316 (2002)
97. W.P. Bowen, N. Treps, B.C. Buchler, R. Schnabel, T.C. Ralph, H.A. Bachor, T. Symul, L.P. Koy, Experimental investigation of continuous-variable quantum teleportation. Phys. Rev. A **67**(3), 032302 (2003)
98. N. Takei, T. Aoki, S. Koike, K. Yoshino, K. Wakui, H. Yonezawa, T. Hiraoka, Experimental demonstration of quantum teleportation of a squeezed state. Phys. Rev. A **72**(4), 042304 (2005)
99. E. Jung, M.R. Hwang, Y.H. Ju, K. MinSoo, K.Y. Sahng, K. Hungsoo, P. DaeKil, S. JinWoo, S. Tamaryan, C. SeongKeuck, Greenberger-Horne-Zeilinger versus W states: quantum teleportation through noisy channels. Phys. Rev. A **78**(1), 012312 (2008)
100. X.M. Jin, J.G. Ren, B. Yang, Z.H. Yi, F. Zhou, X.F. Xu, S.K. Wang, et al., Experimental free-space quantum teleportation. Nat. Photon. **4**(6), 376–381 (2010)
101. M.L. Hu, Environment-induced decay of teleportation fidelity of the one-qubit state. Phys. Lett. A **375**(21), 2140–2143 (2011)
102. H.Q. Liang, J.M. Liu, S.S. Feng, J.G. Chen, Quantum teleportation with partially entangled states via noisy channels. Quantum Inf. Process. **12**(8), 2671–2687 (2013)
103. K.P. Seshadreesan, J.P. Dowling, G.S. Agarwal, Non-Gaussian entangled states and quantum teleportation of Schrödinger-cat states. Phys. Scr. **90**(7), 074029 (2015)
104. R. Fortes, G. Rigolin, Fighting noise with noise in realistic quantum teleportation. Phys. Rev. A **92**(1), 012338 (2015)
105. T.M. Graham, H.J. Bernstein, T.C. Wei, M. Junge, G.K. Paul, Superdense teleportation using hyperentangled photons. Nat. Commun. **6**(7158), 1–21 (2015)

106. M. Zuppardo, T. Krisnanda, T. Paterek, S. Bandyopadhyay, A. Banerjee, P. Deb, S. Halder, K. Modi, M. Paternostro, Excessive distribution of quantum entanglement. Phys. Rev. A **93**(1), 012305 (2016)
107. X. Xing, Y. Yao, W.J. Zhong, Y.L. Li, Y.M. Xie, Enhancing teleportation of quantum Fisher information by partial measurements. Phys. Rev. A **93**(1), 012307 (2016)
108. T. Li, Z.Q. Yin, Quantum superposition, entanglement, and state teleportation of a microorganism on an electromechanical oscillator. Sci. Bull. **61**(2), 163–171 (2016)

Chapter 2
Theoretical Basis

To understand how quantum teleportation protocol works, and to better explain our research work, this chapter introduces the basic theoretical knowledge of quantum entanglement-based teleportation protocol, including the necessary mathematical basis, the basic assumptions and principles of quantum mechanics, the density matrix, the principle of quantum entanglement and entanglement exchange, and the quantum noise characteristics of open systems, which lays a good foundation for the subsequent chapters of this book.

2.1 Fundamentals of Mathematics

The main research objects of linear algebra are vectors and vector spaces. Mastering linear algebra is the basis of understanding quantum mechanics.

2.1.1 Linear Algebra and Vector Space

For any isolated physical system, there is a system state space, which we call Hilbert space in linear algebra. This space is the set of possible states of the whole system, and it is a complex vector space with the inner product defined. In linear algebra, a vector space that defines an inner product is called an inner product space. When an inner product space has completeness (if every Cauchy sequence in an inner product space H converges to an element X in H according to modulus, then H is complete), it is called a Hilbert space.

Quantum bit is the basis of quantum information, and its position in quantum information theory is self-evident. We can use a vector in two-dimensional Hilbert space to represent the state of quantum bits. This vector is called the state vector, the symbol is $|\cdot\rangle$ (bra), and its dual vector is $\langle\cdot|$(ket). In an n-dimensional vector space,

D. Li, *Quantum Teleportation Under Noise*,
https://doi.org/10.1007/978-3-031-91250-4_2

if there a set of vectors $|a_1\rangle$, …, $|a_n\rangle$ can be expressed as $|a\rangle = \sum_i c_i|a_i\rangle$ (where c_i are scalar coefficients), then this set of vectors is called a generating set. If for any list of n complex numbers $c_1c_2c_3\ldots.\ c_n$ we have that $c_1|a_1\rangle + c_2|a_2\rangle + \ldots\ldots + c_n|a_n\rangle = 0$ where $c_1, c_2, c_3, \ldots, c_n$ is not all 0, then the set of vectors $a_1, a_2, a_3, \ldots, a_n$ is said to be linearly independent. A linearly independent generating set is known as a basis. We can use such a basis to represent any quantum state in the system. Generally, a vector space may have many different bases.

For example, a set of bases of vector space C^2 are

$$|v_1\rangle = \begin{bmatrix} 1 \\ 0 \end{bmatrix}, |v_2\rangle = \begin{bmatrix} 0 \\ 1 \end{bmatrix}. \tag{2.1}$$

Then any vector $|\,V\,\rangle$ in C^2 can be written as

$$|v\rangle = \begin{bmatrix} a_1 \\ a_2 \end{bmatrix} = a_1|v_1\,\rangle + a_2|v_2\,\rangle\,. \tag{2.2}$$

Another set of bases can also be used to represent

$$|v\rangle = \begin{bmatrix} a_1 \\ a_2 \end{bmatrix} = \frac{a_1 + a_2}{2}|v_3\,\rangle + \frac{a_1 - a_2}{2}|v_4\,\rangle \tag{2.3}$$
$$|v_3\rangle = \frac{1}{\sqrt{2}}\begin{bmatrix} 1 \\ 1 \end{bmatrix}, |v_4\rangle = \frac{1}{\sqrt{2}}\begin{bmatrix} 1 \\ -1 \end{bmatrix}$$

2.1.2 Vector Correlation Operation

2.1.2.1 Inner Product

In the previous section, the vector space that defines the inner product and satisfies the completeness is called Hilbert space, where the inner product is a binary complex function on the vector space. We can select any two states in the state vector space $|\alpha\rangle$ and $|\beta\rangle$, under a standard orthogonal basis $|v_i\rangle$, and define their inner product as

$$\begin{aligned} c = \langle\alpha|\beta\,\rangle &= \left(\sum\nolimits_i a_i^*\langle v_i|\right)\left(\sum\nolimits_i b_i|v_i\rangle\right) = \sum\nolimits_i a_i^* b_i\langle v_i|v_i\rangle = \sum\nolimits_i a_i^* b_i \\ &= \left[a_1^*\ldots a_i^*\right]\begin{bmatrix} b_1 \\ \ldots \\ b_i \end{bmatrix} \end{aligned} \tag{2.4}$$

where $\langle\alpha|$ represents the dual vector of $|\alpha\rangle$ and a_j^* indicates the complex conjugation of corresponding components after decomposition under the standard orthogonal basis $|v_i\rangle$. We can see that the result of the inner product operation is a complex number. If the inner product of two vectors is 0, we call two vectors orthogonal. If two unit vectors are dual, the inner product is 1.

2.1.2.2 Module and Normalization

The inner product of the vector $|\alpha\rangle$ is written as $\langle\alpha|\,\alpha\rangle = |\alpha|^2$, which is a number greater than or equal to 0. The modules of vector a is defined as $||\ |\alpha\rangle\ || = \sqrt{\langle\alpha|\alpha\rangle} = \sqrt{\sum_i |a_i|^2}$. If $\|\ |\alpha\rangle\ \| = 1$, $|\alpha\rangle$ is called the unit vector, or a normalized vector.

The vector can be normalized by dividing the vector itself by its modulus

$$|\widehat{\alpha}\rangle = \frac{|\alpha\rangle}{||\ |\alpha\rangle\ ||}. \tag{2.5}$$

In an orthogonal basis, each vector is a unit vector, and different vectors are orthogonal to each other.

2.1.2.3 Cross Product

$|\alpha\rangle\langle\beta|$ represents the dot product of two vectors. Observing the final result of the inner product, we can easily find that the dot product is a matrix. Generally, dot product representation is a useful method to express linear operators by the inner product. Assuming that $|v\rangle$ is a vector on inner product space V and $|W\rangle$ represents a vector on inner product space W, define $|w\rangle\langle v|$ as a linear operator from V to W:

$$(|w\rangle\langle v|)(|v'\rangle) = |w\rangle\langle v|v'\rangle = \langle v|v'\rangle\,|w\rangle. \tag{2.6}$$

2.1.2.4 Tensor Product

The tensor product is a method of expanding vector space. It combines two or more vector spaces to form a larger vector space. Assuming that M and N are Hilbert spaces with dimensions M and N, respectively, $M \otimes N$ represents a vector space with dimension mn, and the elements in $M \otimes N$ are linear combinations of elements in M and N. If a set of standard orthogonal bases of M and N are $|i\rangle$ and $|i\rangle$, then $|i\rangle \otimes\ |\ i\rangle$ is a set of standard orthogonal bases of $M \otimes N$.

For arbitrary scalar $z, m_1, m_2 \in M, n_1, n_2 \in N$, the tensor product has the following basic properties:

$$
\begin{aligned}
&z(|m_1\rangle \otimes |n_1\rangle) = (z|m_1\rangle) \otimes |n_1\rangle = |m_1\rangle \otimes (z|n_1\rangle) \\
&(|m_1\rangle + |m_2\rangle) \otimes |n_1\rangle = |m_1\rangle \otimes |n_1\rangle + |m_2\rangle \otimes |n_1\rangle \\
&|m_1\rangle \otimes (|n_1\rangle + |n_2\rangle) = |n_1\rangle \otimes |m_1\rangle + |n_2\rangle \otimes |m_1\rangle
\end{aligned}
\tag{2.7}
$$

More intuitively, we can observe from the Kronecker product, which is expressed as

$$
\begin{bmatrix} a_1 \\ \vdots \\ \vdots \\ \vdots \\ a_n \end{bmatrix} \otimes \begin{bmatrix} b_1 \\ \vdots \\ \vdots \\ \vdots \\ b_n \end{bmatrix} = \begin{bmatrix} a_1 \times \begin{bmatrix} b_1 \\ \vdots \\ \vdots \\ \vdots \\ b_n \end{bmatrix} \\ a_2 \times \begin{bmatrix} b_1 \\ \vdots \\ \vdots \\ \vdots \\ b_n \end{bmatrix} \\ \vdots \\ a_2 \times \begin{bmatrix} b_1 \\ \vdots \\ \vdots \\ \vdots \\ b_n \end{bmatrix} \end{bmatrix} = \begin{bmatrix} a_1 b_1 \\ \vdots \\ a_1 b_n \\ a_2 b_1 \\ \vdots \\ \vdots \\ a_2 b_n \\ \vdots \\ \vdots \\ \vdots \\ a_n b_n \end{bmatrix}. \tag{2.8}
$$

Generally, for quantum bits, we define matrix representation as follows:

$$
|0\rangle = \begin{bmatrix} 1 \\ 0 \end{bmatrix} \ |1\rangle = \begin{bmatrix} 0 \\ 1 \end{bmatrix}. \tag{2.9}
$$

Then for arbitrary pair of qubits, two particles can form four non-repeating qubit pairs $|00\rangle$, $|01\rangle$, $|10\rangle$, $|11\rangle$, and their tensor products can be expressed in matrix as follows:

$$
\begin{aligned}
&|00\rangle = |0\rangle \otimes |0\rangle = \begin{bmatrix} 1 \\ 0 \end{bmatrix} \otimes \begin{bmatrix} 1 \\ 0 \end{bmatrix} = \begin{bmatrix} 1 \\ 0 \\ 0 \\ 0 \end{bmatrix} \\
&|01\rangle = |0\rangle \otimes |1\rangle = \begin{bmatrix} 1 \\ 0 \end{bmatrix} \otimes \begin{bmatrix} 0 \\ 1 \end{bmatrix} = \begin{bmatrix} 0 \\ 1 \\ 0 \\ 0 \end{bmatrix}
\end{aligned}
\tag{2.10}
$$

$$|10\rangle = |1\rangle \otimes |0\rangle = \begin{bmatrix} 0 \\ 1 \end{bmatrix} \otimes \begin{bmatrix} 1 \\ 0 \end{bmatrix} = \begin{bmatrix} 0 \\ 0 \\ 1 \\ 0 \end{bmatrix}$$

$$|11\rangle = |1\rangle \otimes |1\rangle = \begin{bmatrix} 0 \\ 1 \end{bmatrix} \otimes \begin{bmatrix} 0 \\ 1 \end{bmatrix} = \begin{bmatrix} 0 \\ 0 \\ 0 \\ 1 \end{bmatrix}$$

We can intuitively see the overall state of the quantum bit pair through the tensor product. In a multiparticle system, the overall system state can also be expressed by the tensor product. The expression form of the tensor product of the state vector is $|\phi\rangle \otimes |\psi\rangle$, which generally omits the tensor product symbol, abbreviated as $|\phi\rangle|\psi\rangle$ or $|\psi\rangle|\phi\rangle$.

2.1.2.5 Eigenvectors and Eigenvalues

If there is a linear operator $\widehat{A}$, and $\widehat{A}|\psi\rangle = \lambda|\psi\rangle$, where λ is a constant, then lambda is called eigenvalue. $|\psi\rangle$ is an eigenvector (or eigenstate) of $\widehat{A}$. $\widehat{A}|\psi\rangle = \lambda|\psi\rangle$ is called the eigenvalue equation of $\widehat{A}$. According to the basic theorem of algebra, the roots of the characteristic equation is the eigenvalue of the operator $\widehat{A}$. Each operator $\widehat{A}$ has at least one corresponding eigenvector and eigenvalue. And the trace of the operator $\widehat{A}$ is the sum of its eigenvalues. The eigenspace corresponding to an eigenvalue λ is a set of eigenvectors with λ as the eigenvalue. The concept of eigenspace will be used in the subsequent explanation of operator diagonalization.

2.1.3 Operators and Related Concepts

2.1.3.1 Linear Operators and Matrices

For the two vector spaces V and W, map $|v\rangle \in V$ to $|w\rangle \in W$, which is called the operator $\widehat{A}$; it can be expressed as $|w\rangle = \widehat{A}|v\rangle$.

If $\widehat{A}(a|v\rangle + b|w\rangle) = a\left(\widehat{A}|v\rangle\right) + b\left(\widehat{A}|w\rangle\right)$, we call the operator $\widehat{A}$ a linear operator.

Generally speaking, when a linear operator is defined on a vector space V, then the operator $\widehat{A}$ refers to maps from V to V.

1. Identity operator I_v: $I_v|v\rangle = |v\rangle$.
2. Zero operator, written as 0, acts as $0|v\rangle = 0$.

The basic properties of linear operators are as follows:

1. The sum of two linear operators is also linear.
2. The product of two linear operators is also linear and acts on the corresponding vector in sequence, for example, $\left(\widehat{A}\widehat{B}\right)|v\rangle = \widehat{A}\left(\widehat{B}|v\rangle\right)$. And generally $\widehat{A}\widehat{B} \neq \widehat{B}\widehat{A}$.

We can better understand it through the matrix representation of linear operators. In fact, the two are completely equivalent. In vector space C^n, the matrix of a vector $|v\rangle$ is expressed as an n-dimensional column matrix. Assuming that matrix A is an $m \times n$ *matrix*, and its element is denoted by A_{ij}. When the matrix A multiplies the vector $|v\rangle$, it changes the n-dimensional column matrix into the m-dimensional column matrix, which is equivalent to transferring the vector in C^n to C^m.

As we can see above, the matrix can be regarded as a linear operator, so the matrix expression of the linear operator can also be given. Suppose that a standard orthogonal basis of vector spaces V and W is $\{v_1, v_2, \ldots, v_m\}$ and $\{w_1, w_2, \ldots, w_n\}$, respectively; we define the operator $\widehat{A}$ as a linear operator from vector space V to vector space W. For $1 \leq j \leq m$, $1 \leq i \leq n$, there is a complex number A_{ij} which makes

$$\widehat{A}|v_i\rangle = \sum_i A_{ij}|w_i\rangle. \tag{2.11}$$

Matrix A consists of A_{ij} satisfying the condition and is a matrix representation of the linear operator $\widehat{A}$.

2.1.3.2 Hermite Operator

Assume that $\widehat{A}$ is a linear operator on Hermite space V; then there is a unique linear operator $\widehat{A}^{+}$, such that if $|v\rangle, |w\rangle \in V$, then $\left\langle v\middle|\widehat{A}\middle|w\right\rangle = \left\langle v\middle|\widehat{A}^{+}\middle|w\right\rangle$. Operator $\widehat{A}^{+}$ is the Hermitian conjugate operator or the adjoint of $\widehat{A}$. The matrix expression of $\widehat{A}^{+}$ is A; then the Hermite conjugate operation can be written as $A^{+} = (A^{*})^{T}$, where A^{*} represents complex conjugate and T represents transpose. If the Hermite conjugate of an operator is equal to the original operator, then we call this Hermite operator or self-adjoint operator. Hermite operator has the following important properties:

1. In any state, the eigenvalue of the Hermite operator must be a real number.
2. In any state, the operator whose average value is a real number must be the Hermite operator.
3. Eigenvectors of Hermite operators belonging to different eigenvalues are orthogonal to each other.
4. The eigenvector of the Hermite operator stretches a complete vector space.

2.1.3.3 Unitary Operator

If the operator satisfies the condition, $\widehat{A}\widehat{A}^{+} = \widehat{A}^{+}\widehat{A} = I$. The Hermite conjugate operator of an operator is equal to the inverse operator of it, so we call this operator a unitary operator. The unitary operator has the following properties:

1. If $\widehat{A}$ is unitary operator, then $\widehat{A}^{+}, A^{-1}$ is unitary.
2. The product of two unitary operators is also a unitary operator.
3. The unitary operator can keep the inner product between two vectors unchanged.

2.1.3.4 Normal Operator and Spectral Decomposition

The definition of a normal operator is $\widehat{A}\widehat{A}^{+} = \widehat{A}^{+}\widehat{A}$. We call $\widehat{A}$ a normal operator, and the Hermite operator and the unitary operator are both normal operators.

When an operator $\widehat{A}$ is diagonalizable, $\widehat{A}$ can be expressed as

$$\widehat{A} = \sum\nolimits_{i}^{n} \lambda_i |i\rangle\langle i| = \begin{pmatrix} \lambda_1 & 0 & 0 & 0 \\ 0 & \lambda_2 & 0 & 0 \\ \cdots & \cdots & \cdots & \cdots \\ 0 & 0 & 0 & \lambda_n \end{pmatrix} \tag{2.12}$$

where $|i\rangle$ denotes the eigenvector of $\widehat{A}$, and the equation is also called the spectral decomposition of the operator $\widehat{A}$, which is composed of the eigenvalues of $\widehat{A}$.

Spectral decomposition theorem states that if and only if an operator is normal, it can be diagonalized and has orthogonal normalized eigenvector.

2.1.3.5 Projection Operator

The function of the projection operator is to project one vector to the direction of another vector.

Assuming that the two vectors are, respectively, $|\alpha\rangle$ and $|\beta\rangle$, the projection operator of $|\beta\rangle$ projection to $|\alpha\rangle$ is defined as $P_\alpha = |\alpha\rangle\langle\alpha|$. The vector after projection is $P_\alpha|\beta\rangle = |\alpha\rangle\langle\alpha|\beta\rangle = \langle\alpha|\beta\rangle\,|\alpha\rangle$.

2.1.3.6 Commutator and Anti-Commutator

The commutator is defined as $[A, B] = \widehat{A}\widehat{B} - \widehat{B}\widehat{A}$. When $\widehat{A}$ and $\widehat{B}$ are commutative operators, $\widehat{A}$ and $\widehat{B}$ satisfy the condition $[A, B] = \widehat{A}\widehat{B} - \widehat{B}\widehat{A} = 0$. The anti-commutator is defined as $\{A, B\} = \widehat{A}\widehat{B} + \widehat{B}\widehat{A}$. When $\widehat{A}$ and $\widehat{B}$ are anti-commutative operators, $\widehat{A}$ and $\widehat{B}$ satisfy the condition $\{A, B\} = \widehat{A}\widehat{B} + \widehat{B}\widehat{A} = 0$.

2.2 Common Logic Gates

In the protocol of quantum teleportation, we can describe the changes of the system through the language of quantum computing, with the help of quantum logic gates and logic circuits. Let's introduce some commonly used quantum logic gates.

2.2.1 *Single-Qubit Gates*

According to the state superposition principle (Sect. 2.3.2), a single qubit can be expressed as

$$|v\rangle = a_1|0\rangle + a_2|1\rangle \ \left(a_1^2 + a_2^2 = 1\right). \tag{2.13}$$

Its matrix is expressed as

$$|v\rangle = \begin{bmatrix} a_1 \\ a_2 \end{bmatrix}. \tag{2.14}$$

Then changing the single-qubit state can be understood as changing the form of this matrix. From the basic knowledge of the matrix, we know that to change this matrix, we need to multiply a 2 × 2 square matrix left. At the same time, to meet the normalization condition $\left(a_1^2 + a_2^2 = 1\right)$, we need to use a property of the previous unitary transformation. The unitary change will not change the inner product of the vector, so this 2 × 2 square matrix must be a unitary transformation. Here we list some commonly used qubit logic gates.

2.2.1.1 Hadamard Gate

The matrix form of H-gate is as follows:

$$H = \frac{1}{\sqrt{2}}\begin{bmatrix} 1 & 1 \\ 1 & -1 \end{bmatrix}. \tag{2.15}$$

The function of H-gate is to convert base {0,1} and base {+, −} to each other.

$$H|0\rangle = |+\rangle, H|1\rangle = |-\rangle \tag{2.16}$$

where $|+\rangle = \frac{1}{\sqrt{2}}(|0\rangle + |1\rangle), |-\rangle = \frac{1}{\sqrt{2}}(|0\rangle - |1\rangle)$.

Easy to know by calculation, $H^2|\alpha\rangle = |\alpha\rangle$.

2.2.1.2 Pauli Matrix

The function of the Pauli-X gate is to perform the logical non-operation on quantum bits, which is generally recorded as σ_x. The matrix is expressed as

$$X = \begin{bmatrix} 0 & 1 \\ 1 & 0 \end{bmatrix}. \tag{2.17}$$

After the action of X gate, $X|v\rangle = a_2|0\rangle + a_1 | 1\rangle$.

The function of the Pauli-Y gate is to do logical non-operation on qubits. Under the Bloch sphere, the target qubit is rotated 180° around the Y-axis, which is generally recorded as σ_Y. The matrix is expressed as

$$Y = \begin{bmatrix} 0 & -i \\ i & 0 \end{bmatrix}. \tag{2.18}$$

After the action of Y gate, $Y|v\rangle = e^{\frac{i\pi}{2}}(a_1|1\rangle - a_2|0\rangle)$. We usually also use $-i\sigma_Y$ as a transformation; this writing method will also be easier to understand: $-iY|v\rangle = a_2|0\rangle - a_1 | 1\rangle$. Its function is to change $|1\rangle$ to $|0\rangle$ and $|0\rangle$ to $-|1\rangle$.

The function of the Pauli-Z gate is to phase-transform quantum bits, which is generally recorded as σ_Z. The matrix is expressed as

$$Z = \begin{bmatrix} 1 & 0 \\ 0 & -1 \end{bmatrix}. \tag{2.19}$$

After the action of Z gate, $Z|v\rangle = a_1|0\rangle - a_2|1\rangle$.

The remaining Pauli-I gate is the identity transformation, and the matrix is expressed as the identity matrix.

2.2.1.3 S-Gate

S-gate is a phase gate, which essentially changes the phase of the qubit state. Its function is to operate the target qubit so that its state rotates 90° counterclockwise around the Z-axis of the Bloch sphere. The matrix is expressed as

$$S = \begin{bmatrix} 1 & 0 \\ 0 & i \end{bmatrix}. \tag{2.20}$$

After the action of S-gate, $S|v\rangle = a_1|0\rangle + ia_2|1\rangle$.

2.2.1.4 T Gate

T gate is also a phase gate, which operates the target qubit so that the state of the target qubit rotates 45° counterclockwise around the Z-axis of the Bloch sphere. The matrix is expressed as

$$T = \begin{bmatrix} 1 & 0 \\ 0 & e^{\frac{i\pi}{4}} \end{bmatrix}. \tag{2.21}$$

After the action of S-gate, $T|v\rangle = a_1|0\rangle + e^{\frac{i\pi}{2}}a_2|1\rangle$.

2.2.2 *Multi-Qubit Gates*

2.2.2.1 CNOT-Gate

The CNOT-gate should be the most used double-qubit gate, which requires two inputs, one as the control bit and the other as the target bit. Suppose that the control bit qubit is $|x\rangle$ and the target bit is $|y\rangle$; the implementation circuit is as follows (Fig. 2.1):

Its matrix is expressed as

$$CNOT = \begin{bmatrix} 1 & 0 & 0 & 0 \\ 0 & 1 & 0 & 0 \\ 0 & 0 & 0 & 1 \\ 0 & 0 & 1 & 0 \end{bmatrix}. \tag{2.22}$$

After the action of CNOT-gate, the result is $CNOT|xy\rangle = |x\rangle \mid x \oplus y\rangle$, where $\oplus$ represents modulo two addition operation.

2.2.2.2 SWAP-Gate

SWAP-gate is to exchange the positions of the first qubit and the second qubit. The implementation circuit is as follows (Fig. 2.2):

Its matrix is expressed as

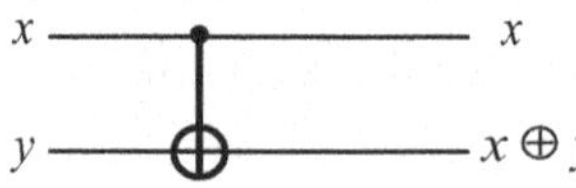

Fig. 2.1 Realization circuit of CNOT-gate

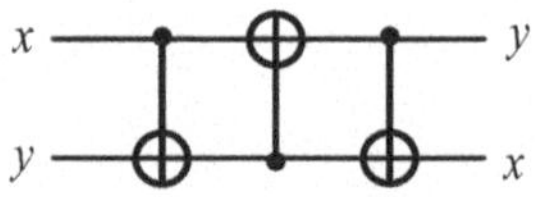

Fig. 2.2 Realization circuit of SWAP-gate

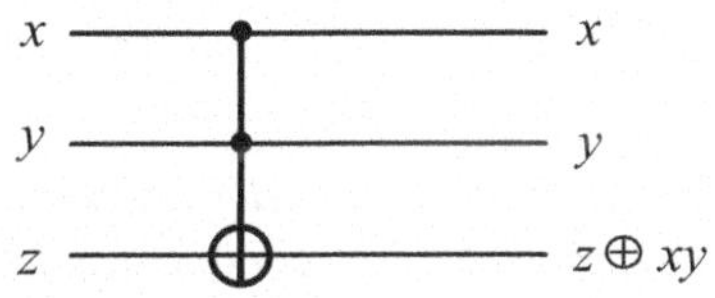

Fig. 2.3 Realization circuit of Toffoli gate

$$SWAP = \begin{bmatrix} 1 & 0 & 0 & 0 \\ 0 & 0 & 1 & 0 \\ 0 & 1 & 0 & 0 \\ 0 & 0 & 0 & 1 \end{bmatrix}. \quad (2.23)$$

2.2.2.3 Toffoli Gate

Toffoli gate is one of the common three-qubit gates. When the first and second qubits are in the quantum state $|1\rangle$, the flip operation is performed on the third qubit. It has two control bits. The implementation circuit of Toffoli gate is as follows (Fig. 2.3):

Its matrix is expressed as

$$\text{Toffoli} = \begin{bmatrix} 1 & 0 & 0 & 0 & 0 & 0 & 0 & 0 \\ 0 & 1 & 0 & 0 & 0 & 0 & 0 & 0 \\ 0 & 0 & 1 & 0 & 0 & 0 & 0 & 0 \\ 0 & 0 & 0 & 1 & 0 & 0 & 0 & 0 \\ 0 & 0 & 0 & 0 & 1 & 0 & 0 & 0 \\ 0 & 0 & 0 & 0 & 0 & 1 & 0 & 0 \\ 0 & 0 & 0 & 0 & 0 & 0 & 0 & 1 \\ 0 & 0 & 0 & 0 & 0 & 0 & 1 & 0 \end{bmatrix}. \quad (2.24)$$

2.3 Basic Principles of Quantum Mechanics

The core theory of quantum teleportation is quantum mechanics. Here are some basic principles of quantum mechanics related.

2.3.1 Uncertainty Principle

The heuristic discussion of this principle was first proposed by W. Heisenberg in 1927. He hoped to successfully qualitatively analyze and express the physical properties of simple quantum experiments [1]. The uncertain relation based on commutation was first deduced by H. P. Robertson in 1929 [2]. The expression in [3] is quoted here. The uncertainty relation of two observable quantities A and B is

$$\triangle(A)\triangle(B) \geq \frac{1}{2}|\overline{C}| \tag{2.25}$$

where $\triangle(A)$ and $\triangle(B)$, respectively, represent the uncertainty (root mean square deviation) of observable quantities A and B in a given quantum state where $\triangle(A) = \sqrt{\langle\psi|A^2|\psi\rangle - \langle\psi|A|\psi\rangle^2}$, $\triangle(B) = \sqrt{\langle\psi|B^2|\psi\rangle - \langle\psi|B|\psi\rangle^2}$, C is the Hermite commutation of A and B, and $C = i[B, A], \overline{C} = \langle\psi|C|\psi\rangle$.

According to the uncertainty principle, in a given quantum state, two anti-commutative observable quantities cannot have definite values at the same time. But there are also special cases. If the quantum state satisfies

$$\overline{C} = \langle\psi|C|\psi\rangle = 0, \tag{2.26}$$

observable quantities A and B can have definite values at the same time [3].

2.3.2 *Principle of Superposition of States*

The superposition principle in mathematics shows that the linear superposition of any solution of the linear equation is still the solution of the original equation. Because the Schrödinger equation is a linear equation, it also satisfies the superposition principle, which is called the superposition principle of states in quantum mechanics [4]. Assuming that mutually orthogonal quantum states are possible states of quantum systems, their superposition states can be expressed as

$$|\psi\rangle = \sum_i c_i|\phi_i\rangle \tag{2.27}$$

where $c_i = \langle\varphi_i \mid \psi\rangle$. When measuring the superposition state $|\psi\rangle$, c_i represents the probability amplitude of getting the orthogonal state $|\varphi_i\rangle$ and satisfies $\sum_i |c_i|^2 = 1$.

The superposition of quantum states is a very important property in quantum informatics. The quantum parallelism and coherence of quantum states in quantum computing all depend on the superposition of quantum states.

2.3.3 *Quantum No-Cloning Theorem*

Based on the superposition principle of quantum states, W. K. Wootters and W. H. Zurek proposed the quantum non-cloning theorem in 1982 [5], that is, an unknown quantum state cannot be cloned completely and accurately. We take the polarization state of photons as an example, and the proof in [6] is cited here. Suppose $|\psi\rangle$ is an unknown quantum state, assuming that there is a physical process U that can be cloned completely and accurately.

$$U(|\psi\rangle|0\rangle)\rightarrow|\psi\rangle|\psi\rangle. \tag{2.28}$$

This process is independent of quantum state $|\psi\rangle$. For any quantum state $|\varphi\rangle\neq|\psi\rangle$, there are also

$$U(|\phi\rangle|0\rangle)\rightarrow|\phi\rangle|\phi\rangle. \tag{2.29}$$

For quantum state $|\varphi\rangle=\frac{1}{\sqrt{2}}(|\psi\rangle+|\phi\rangle)$,

$$U(|\varphi\rangle|0\rangle)=U\left(\frac{1}{\sqrt{2}}(|\psi\rangle+|\phi\rangle)|0\rangle\right)\rightarrow\frac{1}{\sqrt{2}}(|\psi\rangle|\psi\rangle+|\phi\rangle|\phi\rangle)\neq|\varphi\rangle|\varphi\rangle. \tag{2.30}$$

The result is not the cloning of quantum state $|\phi\rangle$, so the hypothesis is not tenable, and such a physical process does not exist.

Although the quantum non-cloning theorem proves that unknown quantum states cannot be accurately cloned, quantum states can be cloned in some special cases. H. P. Yuen proved [7] if a unitary transformation is used to represent the replication process, if and only if two quantum states are orthogonal, they can be cloned by the same replication process. The previous two conclusions are for pure states. For mixed states, H. Barnum et al. [8] proposed the no broadcasting theory.

2.4 Density Matrix

In quantum mechanics, superposition states are different from mixed states. Superposition states belong to pure states and can be directly described by wave functions or vectors, but mixed states cannot be described only by a density matrix. The density matrix is a very useful tool in quantum mechanics. It can bring more information than the state vector.

Suppose that the probability of a system in the state $|\psi_i\rangle$ is P_i; we call $\{P_i,|\psi_i\rangle\}$ an ensemble of pure states. The density matrix of the system is defined as follows:

$$\rho=\sum_i P_i|\psi_i\rangle\langle\psi_i|. \tag{2.31}$$

When the system is a pure state, P_i only has a unique value and $P_i=1$; then the pure state can be represented by the density matrix $\rho=|\psi\rangle\langle\psi|$.

The properties of the density matrix are as follows:

1. The trace of the density matrix is 1.
2. The density matrix is a Hermite matrix.

3. The density matrix is a positive semidefinite matrix.
4. For pure states, $tr(\rho^2) = tr\,\rho = 1$. For mixed states, $tr(\rho^2) < 1$.

We don't always study single systems. When the system we study is a subsystem of a composite system, we need to use the reduced density matrix. Assuming that the subsystem is A and the external system is B, the state of the whole system can use the density matrix ρ^{AB} to express; we define the reduced density matrix of the subsystem as $\rho^A = tr_B(\rho^{AB})$, where tr_B is an operator map, which is called the partial trace on system B.

Suppose the two vectors in system A are a_1, a_2. The two vectors in system B are b_1, b_2. We can define the partial trace as follows:

$$tr_B(\,|a_1\rangle\langle a_2\,| \otimes |b_1\rangle\langle b_2| = |a_1\rangle\langle a_2\,|\,tr_B(|b_1\rangle\langle b_2\,|) \tag{2.32}$$

where $tr_B(|b_1\rangle\langle b_2\,|) = \langle b_2 \mid b_1\rangle$.

2.5 Quantum Entanglement and Entanglement Swapping

The properties of particles in the system will become the properties of the whole composite physical system, and they can no longer describe the properties of each subsystem alone. This phenomenon is called quantum entanglement [9].

The phenomenon of quantum entanglement was first described by A. Einstein, B. Podolsky, and N. Rosen in 1935 [10]. They discussed the entangled state of the two-free-particle system and proposed the famous EPR paradox. The term "entanglement" was first given by E. Schrödinger [11]. In the 1950s, D. Bohm proposed the implicit variable theory, and in his book [12], taking the two-particle system composed of spin 1/2 particles as an example, the EPR paradox is discussed. In the decades after the proposal of quantum entanglement, the party led by Einstein and the other party led by Bohr launched a fierce debate on the completeness of quantum mechanics, but the debate was limited to the scope of epistemology and philosophy [13]. In 1964, based on the local implicit variable theory, J. S. Bell analyzed the Bohm form of the EPR paradox and obtained the famous Bell inequality [14]. In 1969, J. F. Clauser et al. analyzed the correlation between the spin components of two particles in four different directions and obtained CHSH inequality similar to Bell inequality [15]. Compared with Bell inequality, CHSH inequality is more practical for experimental comparison. In 1982, A. Aspect et al. [16] verified experimentally for the first time that CHSH inequality was violated in the quantum world, and the experimental results were consistent with the expectations of quantum mechanics. Subsequently, a large number of experiments also proved that CHSH inequality was violated [17]. In the field of quantum cryptography, quantum entanglement is used as a new resource to complete the difficult tasks in the classical world.

2.5.1 Quantum Entanglement

Although the concept of quantum entanglement was proposed in 1935, it was not until 1989 that R. F. Werner gave a more complete description of quantum entanglement mathematically for the first time [18]. Due to the complexity of Werner's description, the definition in [19] is cited here. Firstly, the definition of entanglement of two-body pure states is given.

For a two-body pure state $|\psi\rangle \in H_{AB}$, if there are two quantum states $|\phi_A\rangle \in H_A$ and $|\phi_B\rangle \in H_B$,

$$|\psi\rangle = |\phi^A\rangle \otimes |\phi^B\rangle. \tag{2.33}$$

The quantum state $|\psi\rangle$ is a direct product state or separable. Otherwise, the quantum state is entangled.

If a quantum system is in a direct product state, the quantum subsystems that make up it are independent of each other, and this quantum can be prepared separately. However, quantum entangled states cannot be prepared separately, and they have many strange and interesting properties. Take the following Bell state as an example.

$$|\psi^+\rangle = |\beta_{01}\rangle = \frac{1}{\sqrt{2}}(|01\rangle + |10\rangle). \tag{2.34}$$

When the composite system composed of two particles A and B is in the quantum state, the system has the following properties [20]:

1. The subsystem of particles A and B is in a completely mixed state.
2. When measuring particle A, if the measured state of particle A is $|1\rangle$, then particle B must be in the $|0\rangle$. When the state of particle A is $|0\rangle$, particle B must be in state $|1\rangle$. So particles A and B are always in the opposite state.

2.5.2 Definition of Entanglement of Mixed States

Using the same idea as in the pure state, according to whether a mixed state can be prepared separately, the definitions of separability and entanglement of mixed states can be given [18, 19].

Let ρ be a density operator conforming to the system; if there are quantum states ρ_A in A system and ρ_B in B system, then

$$\rho = \rho^A \otimes \rho^B. \tag{2.35}$$

At this time, quantum state ρ is called a direct product state. If there is a set of weights $p_i(\sum_i p_i = 1)$ and a set of direct product states $\rho_i^A \otimes \rho_i^B$, they meet the condition

$$\rho = \sum_i p_i \rho_i^A \otimes \rho_i^B. \tag{2.36}$$

Then quantum state ρ is separable; otherwise, it is entangled.

Physically speaking, the above definition describes three distinct situations for systems A and B. First, direct product states represent unrelated quantum systems where A and B each possess a separate, independent quantum state. Second, separable states represent classically correlated systems; these are non-direct product states where Alice (owner of A) and Bob (owner of B) can prepare the correlation by negotiating the weights $\mathrm{p_i}$ for each possible outcome i through classical communication. Finally, entangled states (or non-separable states) represent the third situation: they are non-direct product states that cannot be described as separable states and exhibit uniquely quantum correlations beyond classical description.

For each result i, they can generate quantum states through local operations $\rho_i^A \otimes \rho_i^B$. So a quantum state is generated $\rho = \sum_i p_i \rho_i^A \otimes \rho_i^B$. This process does not use the theory of quantum mechanics and only involves the concept of classical correlation. However, if a quantum state is entangled, it cannot be generated by the process described above. In this sense, quantum entanglement is a typical feature of quantum mechanics [19].

2.5.3 Definition of Entanglement of Multi-Body Quantum States

Here we first give the definition of separability and entanglement of multi-body quantum pure states and then give the definition of separability and entanglement of mixed states.

Suppose $|\psi\rangle$ is an N-body quantum pure state. If there is a pure state $|\phi_i\rangle$ for the ith subsystem, it meets the condition

$$|\psi\rangle = \otimes_{\mathrm{i=1}}^{\mathrm{N}} |\phi_{\mathrm{i}}\rangle. \tag{2.37}$$

At this time, quantum state $|\psi\rangle$ is completely separable. If a quantum pure state is not completely separable, it contains some form of entanglement. This is discussed in two cases. If there is a division, n subsystems can be divided into m $(1 < m < N)$ parts $P_1, P_2, \cdots, P_m$ and satisfy

$$|\psi\rangle = \otimes_{\mathrm{i=1}}^{\mathrm{m}} |\phi \mathrm{P_i}\rangle. \tag{2.38}$$

At this time, the quantum state $|\psi\rangle$ is m-separable. If a quantum state is neither completely separable nor m-separable, it is said to be truly entangled. For a mixed state ρ, if it can be written in the form of convex combinations of completely separable quantum pure states, that is,

$$\rho = \sum_k p_k \rho_k^1 \otimes \rho_k^2 \otimes \cdots \otimes \rho_k^N. \tag{2.39}$$

At this time, the mixed state ρ is completely separable. Similarly, if a mixed state can be written as a convex combination of m-separable pure states, the mixed state is said to be m-separable.

2.5.4 *Entanglement Swapping*

Entanglement swapping is a very important property of quantum entanglement, which can establish an entanglement relationship between two quantum systems that are far away from each other and are in independent entanglement sources. This is a unique quantum effect, and there is no corresponding theory in the classical world.

There are two pairs of entangled particles: Particle 1 and particle 2 share the Bell states $|\phi^+\rangle_{12}$. Particle 3 and particle 4 share the Bell states $|\phi^+\rangle_{34}$. They form a composite system $|\phi^+\rangle_{12} \otimes |\phi^+\rangle_{34}$; it can be rewritten as

$$|\phi^+\rangle_{12}|\phi^+\rangle_{34} = \frac{1}{2} \times \left(|\phi^+\rangle_{23}|\phi^+\rangle_{14} + |\phi^-\rangle_{23}|\phi^-\rangle_{14} + |\psi^+\rangle_{23}|\psi^+\rangle_{14} + |\psi^-\rangle_{23}|\psi^-\rangle_{14}\right). \tag{2.40}$$

If the joint Bell state measurement is carried out for particle 2 and particle 3, the four particles will collapse to $|\phi^+\rangle_{23}|\phi^+\rangle_{14}$, $|\phi^-\rangle_{23}|\phi^-\rangle_{14}$, $|\psi^+\rangle_{23}|\psi^+\rangle_{14}$ or $|\psi^-\rangle_{23}|\psi^-\rangle_{14}$. We can see that particle 1 and particle 4 are entangled, and particle 2 and particle 3 are entangled, thus realizing the entanglement swapping.

If the initial particles are in the other Bell states, we get similar results. In addition, the entanglement swapping between two pairs of Bell states can also be extended to the entanglement swapping between multiple pairs of Bell states.

2.6 Quantum Noise Characteristics of Open Systems

Noise has become a decisive factor affecting the safety and accuracy of quantum dialogue. Most practical quantum channels are optical fibers, which have birefringence fluctuations. Because the time window of photon travel is shorter than that of the noise source, photons are often affected by noise when they are transmitted in quantum channels.

2.6.1 Open Systems

For a closed quantum system [21, 22], the Hamiltonian of the system is H; then the evolution of the quantum state of the system is unitary.

$$|\Psi 0\rangle \overset{U(t)}{\rightarrow} |\Psi(0)\rangle = U(t)\,|\Psi(0)\rangle \quad U^{\dagger}(t)U(t) = 1. \tag{2.41}$$

If H does not contain time, then $U(t) = e^{-iHt/h}$. But for an open quantum system, the above Equation 2.41 is no longer applicable, and new methods need to be applied to discuss its evolution.

Suppose that system A and environment B form a closed system, a total system. The Hamiltonians of system and environment are HA and HE, respectively, and their density operators at the initial time are ρ_A and $|0\rangle_E\langle 0|$, respectively. The evolution process of the total system is unitary evolution, which can be expressed as

$$\rho_A \otimes |0\rangle_E\langle 0| \overset{U_{AE}(t)}{\rightarrow} U_{AE}(t)(\rho_A \otimes |0\rangle_E\langle 0|)U^{\dagger}_{AE}(t) \tag{2.42}$$

where $U_{AE}(t)$ is the unitary operator of the total system evolution. Let there be a set of orthogonal complete bases $\{|\mu\rangle_E\}$ in Hilbert space of environment; by tracing the environment, the evolution of the system can be expressed as

$$\begin{aligned}\rho_A &\overset{\$}{\rightarrow} \rho'_A = \mathrm{Tr}_E\left[U_A E(t)(\rho_A \otimes |0\rangle_E\langle 0|)U^{\dagger}_{AE}(t)\right] \\ &= \sum_{\mu} \langle \mu|U_{AE}(t)|0\rangle_E \rho_A \left\langle 0\middle|U^{\dagger}_{AE}(t)\middle|\mu\right\rangle_E) = \sum_{\mu} M_\mu \rho_A M^{\dagger}_\mu\end{aligned} \tag{2.43}$$

where $M_\mu = {}_E\langle \mu|U_{AE}|0\rangle_E$, and it satisfies

$$\sum_{\mu} M^{\dagger}_\mu M_\mu = \sum_{\mu} \left\langle 0\middle|U^{\dagger}_{AE}\middle|\mu\right\rangle_{EE} \langle \mu|U_{AE}|0\rangle_E = {}_E\left\langle 0\middle|U^{\dagger}_{AE}U_{AE}\middle|0\right\rangle_E = I_A. \tag{2.44}$$

M_μ is called the Kraus operator, and the mapping \$ is called the super operator.

2.6.2 Measuring Method

How to extract quantum state information from open systems? This requires measurement of the system. Quantum measurement is described by a set of measurement operators, which satisfy the completeness equation.

$$\sum_{m} M_{m}^{\dagger} M_{m} = I. \tag{2.45}$$

The index m represents the possible measurement results. Let the state of the quantum system before measurement be $|\psi\rangle$; the probability of occurrence of result m is determined by Equation 2.46.

$$p(m) = \langle\psi|M_{m}^{\dagger} M_{m}|\psi\rangle. \tag{2.46}$$

The status of the system after the measurement is as follows:

$$\frac{M_{m}|\psi\rangle}{\sqrt{\langle\psi|M_{m}^{\dagger} M_{m}|\psi\rangle}}. \tag{2.47}$$

The completeness relation states that for any state vector $|\psi\rangle$, the sum of the probabilities of all measured results equals 1:

$$\sum_{m} p(m) = \sum_{m} \langle\psi|M_{m}^{\dagger} M_{m}|\psi\rangle = 1. \tag{2.48}$$

The measurements described above are called general measurements. Two special cases of hypothesis 3, projection measurement and positive operator-valued measure (POVM), are commonly used in quantum cryptography.

2.6.2.1 Projection Measurement

Projection measurement can be characterized by an observable Hermite operator M in the measured state space. M has spectral decomposition form: $M = \sum_{m} mP_{m}$, where m is the eigenvalue of observable M and P_m is the projection subspace corresponding to the eigenvalue m. If the eigenvector corresponding to m is $|\xi m\rangle$, then $P_m = | \xi m\rangle\langle\xi m|$. The possible result of measurement corresponds to the eigenvalue m. If the state of the quantum system before a measurement is $|\psi\rangle$, the probability of measuring the result m is

$$p(m) = \langle\psi|P_{m}|\psi\rangle. \tag{2.49}$$

The status of the system after measurement is

$$\frac{P_{m}|\psi\rangle}{\sqrt{p(m)}}. \tag{2.50}$$

In fact, when considering other axioms of quantum mechanics, projection measurement plus unitary operation is completely equivalent to general measurement.

2.6.2.2 POVM

In some applications, POVM is particularly applicable if we only pay attention to the probability of obtaining different results after measuring the system, and do not care about the state of the system after measurement.

POVM is described by a set of positive semidefinite operators $\{E_m\}$ called POVM elements. Similarly, this set of operators satisfies the completeness equation.

$$\sum_m E_m = I. \tag{2.51}$$

In system state $|\psi\rangle$, the probability of the result m obtained from the above measurement is

$$p(m) = \langle\psi|E_m|\psi\rangle. \tag{2.52}$$

In general measurement, if we defined $E_m \equiv M_m^\dagger M_m$, POVM can be obtained.

2.6.3 *Quantum Fidelity*

In quantum communication, due to the influence of noise and other factors, quantum states may change in the process of transmission. In order to describe the proximity between the two quantum states before and after transmission, the concept of quantum fidelity needs to be introduced.

The fidelity of two quantum states ρ and σ is defined as follows [23]:

$$F(\rho,\sigma) \equiv \mathrm{tr}\sqrt{\rho^{1/2}\sigma\rho^{1/2}}. \tag{2.53}$$

Quantum fidelity is used to measure the closeness between two quantum states, if the value of fidelity is closer to 1. It shows that the closer these two quantum states are. The fidelity of quantum states in two special cases is given below.

If the density matrices of two quantum states ρ and σ are commutative, indicating that they can be diagonalized at the same time, and under the same set of standard orthogonal bases $|i\rangle$, these two density matrices can be expressed as

$$\rho = \sum_i r_i |i\rangle\langle i|, \sigma = \sum_i s_i|i\rangle\langle i|. \tag{2.54}$$

Their fidelity is

$$F(\rho,\sigma)=\mathrm{tr}\sqrt{\sum_i r_i s_i \mid i\rangle\langle i \mid}=\mathrm{tr}\left(\sum_i \sqrt{r_i s_i}|i\rangle\langle i|\right)=\sum_i \sqrt{r_i s_i}=F(r_i,s_i). \quad (2.55)$$

At this time, the fidelity between these two quantum states is reduced to classical fidelity, with the eigenvalues of their density matrix (r_i and s_i) as the probability distributions. Another special case is to calculate the fidelity between pure state $|\psi\rangle$ and quantum state ρ.

$$F(|\psi\rangle,\rho)=\mathrm{tr}\sqrt{\langle\psi\rho \mid \psi\rangle \mid \psi\rangle\langle\psi \mid}=\sqrt{(\langle\psi|\rho|\psi\rangle)}. \quad (2.56)$$

The fidelity between two pure quantum states $|\psi\rangle$ and $|\phi\rangle$ is the modules of their inner product, given by $|\langle\psi \mid \varphi\rangle|$.

2.7 Summary of This Chapter

In Sect. 2.1, this chapter introduces the mathematical basis involved in the research, mainly the basic concepts of linear algebra. The commonly used single-bit logic gates and multi-bit logic gates are introduced in Sect. 2.2. In Sect. 2.3, three basic principles commonly used in quantum teleportation are introduced, including the state superposition principle, the uncertainty principle, and the non-cloning theorem. The density matrix is introduced in Sect. 2.4. Section 2.5 introduces the basic concepts of quantum entanglement and entanglement exchange, entangled states, and related knowledge of entanglement exchange. Section 2.6 introduces the quantum noise characteristics of open systems, including the evolution process of open systems, measurement methods, and the measurement of quantum fidelity, which lay a theoretical foundation for the relevant schemes proposed in Chaps. 3–6 and guide and support the construction of a high-fidelity entangled quantum teleportation channel framework, the separation of quantum information between the Bell states and arbitrary states in different channels, and the design and application of a controllable multi-degree of freedom quantum teleportation protocol for immune noise.

References

1. J.A. Wheeler, W.H. Zurek, *Quantum Theory and Measurement* (Princeton University Press, Princeton, NJ, 1983), pp. 62–84
2. H.P. Robertson, The uncertainty principle. Phys. Lett. **34**(1), 163–164 (1929)
3. 曾谨言 (J. Zeng), 量子力学(卷II 第五版) (Quantum Mechanics, vol. 2, 5th edn). 北京: 科学出版社 (Science Press, Beijing, 2014), p. 128

4. A.P. French, E.F. Taylor, *An Introduction to Quantum Physics* (CRC Press, Boca Raton, FL, 1979), p. 316
5. W.K. Wootters, W.H. Zurek, A single quantum cannot be cloned. Nature **299**(5886), 802–803 (1982)
6. 杨宇光 (Y. Yang). 量子密码协议的设计和分析 (Design and Analysis of Quantum Cryptography Protocols). 北京: 科学出版社 (Science Press, Beijing, 2013), p. 16
7. H.P. Yuen, Amplification of quantum states and noiseless photon amplifiers. Phys. Lett. A **113**(8), 405–407 (1986)
8. H. Barnum, C.M. Caves, C.A. Fuchs, et al., Noncommuting mixed states cannot be broadcast. Phys. Rev. Lett. **76**(15), 405–407 (1996)
9. F. Laloe, *Do We Really Understand Quantum Mechanics* (Cambridge University Press, New York, 2012), pp. 121–312
10. A. Einstein, B. Podolsky, N. Rosen, Can quantum-mechanical description of physical reality be considered complete? Phys. Rev. **47**(10), 777–780 (1935)
11. E. Schrödinger, Die gegenwärtige situation in der Quantenmechanik. Naturwissenschaften **23**(48), 807–812 (1935)
12. D. Bohm, *Quantum Theory* (Courier Corporation, New York, 1951), pp. 611–623
13. Y.Y. Nie, Y.H. Li, J.C. Liu, M.H. Sang, Quantum Inf. Process. **10**(3), 297–305 (2011)
14. E. Oelker, T. Isogai, J. Miller, M. Tse, L. Barsotti, N. Mavalvala, M. Evans, Audio-band frequency-dependent squeezing for gravitational-wave detectors. Phys. Rev. Lett. **116**(4), 041102 (2016)
15. J.F. Clauser, M.A. Horne, A. Shimony, R.A. Holt, Proposed experiment to test local hidden variable theories. Phys. Rev. Lett. **23**(15), 880–884 (1969)
16. J. Aspect, J. Dalibard, G. Roger, Experimental test of Bell's inequalities using time-varying analyzers. Phys. Rev. Lett. **49**(25), 1804–1807 (1982)
17. A. Aspect, J. Dalibard, G. Roger, Bell's inequality test: more ideal than ever. Nature **398**(6724), 189–190 (1999)
18. R.F. Werner, Quantum states with Einstein-Podolsky-Rosen correlations admitting a hidden variable model. Phys. Rev. A **40**(8), 4277–4281 (1989)
19. G. Tóth, O. Gühne, Entanglement detection. Phys. Rep. **474**(1), 1–75 (2009)
20. 刘坤 (K. Liu), 量子纠缠态 (Quantum Entangled States). 科技信息 (Sci. Technol. Inf.) **2010**(21), 234–235 (2010)
21. 范洪义, 胡利云 (H. Fan, L. Hu), 开放系统量子退相干的纠缠态表象论 (Entangled State Representation Theory of Quantum Decoherence in Open Systems). 上海:上海交通大学出版社(Shanghai Jiao Tong University Press, Shanghai, 2010), pp. 2–11
22. 穆青霞 (Q. Mu), 腔中连续变量纠缠态制备的理论研究 (Theoretical Study on Preparation of Continuous Variable Entangled States in Cavities). 大连:大连理工大学 (Dalian University of Technology, 2010), pp. 31–32
23. M.A. Nielsen, I.L. Chuang, *Quantum Computation and Quantum Information: 10th Anniversary Edition* (Cambridge University Press, New York, 2010), pp. 1–607

Chapter 3
Construction of a Framework for High-Fidelity Entangled Quantum Teleportation Channel

The real physical system is a semi-open system. When the real physical system is coupled with the environment, there is not only quantum state decoherence problem but also entanglement death problem. This is an almost inevitable problem, but it has to be overcome for that is a very big threat to quantum teleportation missions. It is found that the anti-noise research mainly includes the use of entanglement purification, single quantum error rejection to overcome the quantum state decoherence effect, and entanglement death problem. Entanglement purification is to extract the maximum entanglement state from the mixed system, which requires a large amount of quantum resources [1]. Single quantum error rejection requires a small amount of quantum resources, but it only completes error rejection in probability [2, 3]. In short, in order to reduce or eliminate the influence of noise, auxiliary particles are introduced to reduce the probability of quantum state decoherence and entanglement death caused by noise. Since a large amount of quantum resources are consumed in the process of quantum teleportation, the channel capacity is inevitably reduced, which cannot meet the needs of remote quantum teleportation, and the immune model can only solve the problem of local noise in a single isolated system. Therefore, build a quantum entanglement evolution model based on multi-degree of freedom characteristics, characteristics of entanglement evolution, noise, and entanglement measure under the influence of noise to characterize the entanglement evolution models for different quantum noise channel systems in locally independent and locally common modes, finding the cause of entanglement and sudden death and controlling the entanglement evolution of quantum systems. To understand the impact of noise on the entanglement evolution of the system and build a real-time and adaptive dynamic switching immune noise model, so that quantum entanglement can be maintained to the maximum extent in the application process, which will provide strong theoretical support for maintaining the maximum entanglement state in the process of quantum teleportation.

The traditional coding method based on noise channel is in time, space, or frequency of single degree of freedom under a single operation, which is the sender (Alice) and the receiver (Bob) under the control side (Charlie) measurement,

D. Li, *Quantum Teleportation Under Noise*,
https://doi.org/10.1007/978-3-031-91250-4_3

introducing auxiliary particle and constructing unitary transformation matrix, purifying quantum channel, and implementing at a certain probability dense coding and quantum error correcting code. The receiver cannot completely distinguish the transmitted quantum states, and the channel capacity does not reach the theoretical value, which cannot meet the needs of long-distance quantum teleportation. In this book, we use the idea of quantum graph cascade coding, the information security theory, and the quantum random coding theory, to make full use of the orthogonal normality of graph state basis to tree graph and forest graph to diagonalize large matrix or block diagonalization and then carry out quantum coding operation. Through the calculation and analysis of different level link code channel capacity can obtain the computational formula of multichannel coherent information of different noise channels under the coded multiple degrees of freedom, quickly calculate the various cascade codes correlation information in the channel, the approximate value of channel capacity and the quantum information noise tolerance on the channel transmission, and analyze the different anti-noise performance of different cascade code in the different parameters of the channel which can obtain the practicable quantum information regions to transmitted in different noise channels and provide a basis for constructing channel adaptation codes.

To construct a unified high-fidelity entangled quantum transmission channel system, the research begins with investigating the entanglement evolution model, robustness against noise models, and channel capacity coding under noise. Specifically, a new quantum entanglement evolution model is proposed, which consists of three components: characterization of entanglement death, characterization of quantum decoherence, and quantum entanglement evolution under local common noise. By analyzing the evolution of entanglement of two mutual independence atomic systems and two three-level atomic systems on the space with time to design the related parameters when entanglement death is happening. The properties of quantum decoherence in the interaction between system and environment can be revealed by analyzing characterization of the quantum decoherence between atom and cavity field in Tavis-Cummings (T-C) model, which lays a foundation for the construction of immune noise model. Secondly, the effects of different noises on the evolution characteristics of entangled mixed state in the local common mode are studied, especially when the quantum channel is affected by the local common quantum noise environment can describe the process of entanglement sudden death and resurrection very accurately, so that an immune model based on density matrix and DFS combined noise could be constructed. Finally, a method based on graph state theory is proposed to apply quantum graph state coding to tree graphs and forest graphs with multiple degrees of freedom. This involves analyzing coherence information and channel capacity, thereby enabling the acquisition of multichannel coherent information under this coding scheme across different noise channels. Additionally, it allows for the effective computation of approximate values of quantum capacity in noisy channels, which aids in calculating the valuable transmission of quantum information while improving the computing speed and noise tolerance of channel-transmitted quantum information. Thus, the region that can transmit quantum information in different noise channels including several

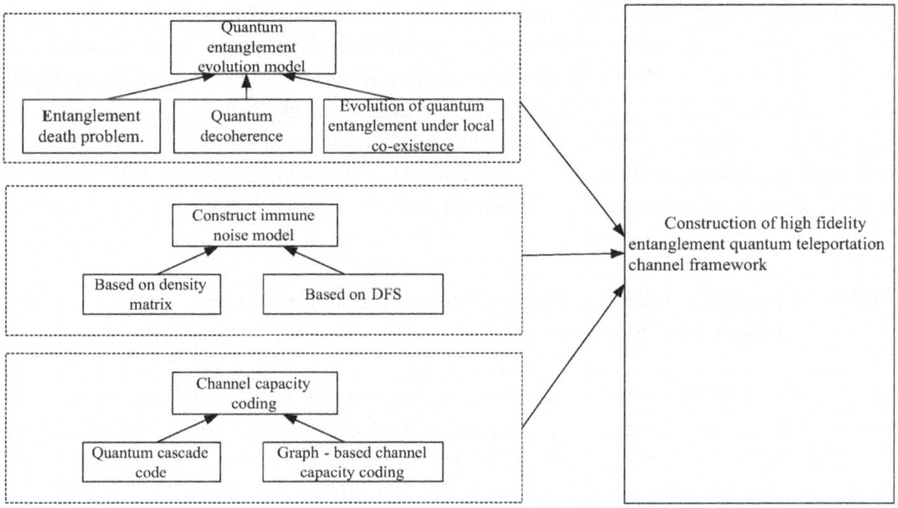

Fig. 3.1 Organizational structure of high-fidelity entanglement quantum teleportation channel framework

common ones can be obtained, which can increase channel capacity effectively. Figure 3.1 shows the organizational structure of high-fidelity entanglement quantum teleportation channel framework.

3.1 Establishment of the Quantum Entanglement Evolution Model

Based on the entanglement characteristics generated by the coupling between the external environment and the quantum state of the system, the study of the sudden death of quantum entanglement and quantum decoherence and the analysis of the influence of different noises on the entanglement evolution of the entangled mixed state in the locally independent and locally common modes are the core of the construction of the entanglement evolution model in open quantum system.

3.1.1 Depiction of Entanglement Death

The real physical system is a semi-open system. When the physical system is coupled with the external environment, it will lead to entanglement between the states in the system and the states in the environment. The quantum system is no longer an ideal vacuum system, and the coupling evolution between each other is no longer unitarian and reversible, and it also no longer follows the Schrödinger equation. The coherence of the system gradually decays and cannot be maintained. At the same time, quantum entanglement associated with quantum correlation will also decay or even suddenly disappear in a phenomenon called entanglement death, thus reducing the quality of service of quantum teleportation. In this section, we describe the evolution characteristics of entanglement over time from two aspects, find out the cause of entanglement death, and lay a foundation for the establishment of the quantum entanglement evolution model.

3.1.1.1 Characterization of Entanglement Evolution of Two Independent Systems

We assume that there are an atom Ω_1 in a two-mode cavity [4] and an isolated atom Ω_2 outside the cavity. If the interaction time between the atoms and the cavity is far less than the maximum lifetime of the cavity (the decoherence effect caused by the cavity consumption can be ignored), the Hamiltonian of the atom and the field can be written in the rotation wave approximation

$$|H\rangle = \sigma_+ ag + \sigma_- a^+ g = (\sigma_+ a + \sigma_- a^+)g \ (\hbar = 1) \tag{3.1}$$

where σ_+ and σ_- are the Pauli spin operator, a^+ is the generation operator of cavity field, a is the annihilation operator of atom, and g is the coupling constant of atom and cavity field.

The atom in the initial entangled state is described as

$$|\Theta(0)\rangle = \cos\theta|e_1g_2\rangle + \sin\theta|g_1g_2\rangle. \tag{3.2}$$

The entangled state of the initial two-mode cavity field is described as

$$|\Theta(0)\rangle_f = \cos\phi\,|10\rangle_{XY} + \sin\phi\,|01\rangle_{XY} \tag{3.3}$$

where XY are two cavity modes, respectively. Thus, the initial state vector of the system composed of XY, Ω_1, and Ω_2 is

$$\begin{aligned} |\Theta(0)\rangle_{\Omega_1\Omega_2 f} &= \cos\theta|e_1g_2\rangle\cos\phi\,|10\rangle_{XY} + \cos\theta|e_1g_2\rangle\sin\phi\,|01\rangle_{XY} \\ &\quad \sin\theta|g_1g_2\rangle\cos\phi\,|10\rangle_{XY} + \sin\theta|g_1g_2\rangle\sin\phi\,|01\rangle_{XY} \\ &= \cos\theta\cos\phi|e_1g_21_X0_Y\rangle + \cos\theta\sin\phi|e_1g_20_X1_Y\rangle \\ &\quad + \sin\theta\cos\phi|g_1g_21_X0_Y\rangle + \sin\theta\sin\phi|g_1g_20_X1_Y\rangle \end{aligned}. \tag{3.4}$$

The atom Ω_1 in the two-mode cavity interacts with the cavity mode, and the state vector obtained by solving the Schrödinger equation:

$$\begin{aligned}|\Theta(0)\rangle_{\Omega_1\Omega_2 f} &= \cos\theta\cos\phi\cos\left(\sqrt{2}gt\right)|e_1g_21_X0_Y\rangle + \cos\theta\sin\phi\cos(gt)|e_1g_21_X0_Y\rangle \\ &+\sin\theta\cos\phi\cos(gt)|g_1e_21_X0_Y\rangle + \sin\theta\sin\phi|g_1e_21_X0_Y\rangle \\ &-i\cos\theta\cos\phi\sin\left(\sqrt{2}gt\right)|e_1g_21_X0_Y\rangle - i\cos\theta\sin\phi\sin(gt)|e_1g_21_X1_Y\rangle \\ &-i\sin\theta\cos\phi\sin(gt)|e_1e_20_X0_Y\rangle\end{aligned} \quad . \tag{3.5}$$

Trace the degree of freedom of the field in the system density matrix $\rho_{\Omega_1\Omega_2 f(t)} = |\Theta(t)_{\Omega_1\Omega_2 f}\rangle\langle\Theta(t)_{\Omega_1\Omega_2 f}|$, the reduced density matrix $\rho_{\Omega\Omega_1\Omega_2(t)}$ of atom Ω_1 and Ω_2 can be obtained, and we describe the expression as

$$\rho_{\Omega_1\Omega_2}(t) = \begin{bmatrix} \rho_{11}(t) & 0 & 0 & 0 \\ 0 & \rho_{22}(t) & \rho_{23}(t) & 0 \\ 0 & \rho_{32}(t) & \rho_{33}(t) & 0 \\ 0 & 0 & 0 & \rho_{44}(t) \end{bmatrix} \tag{3.6}$$

Therefore, we get the entangled information of two atoms Ω_1 and Ω_2 because the entanglement degree of a pure state quantum system can be expressed by the von Neumann entropy of the reduced density matrix of any subsystem.

Entanglement between two atoms can be measured by the degree of concurrency, which is defined as

$$C(t) = m\ [0, \lambda_1(t) - \lambda_2(t) - \lambda_3(t) - \lambda_4(t)] \tag{3.7}$$

where the descending order of eigenvalues of operator $\left[\sqrt{\rho}, \overline{\rho}\sqrt{\rho}\right]^{\frac{1}{2}}$ is $\lambda_1(t)$, $\lambda_2(t)$, $\lambda_3(t)$ and $\lambda_4(t)$. The time inversion matrix $\overline{\rho}, \rho$ is expressed as

$$\overline{\rho} = \sigma_y \bigotimes \sigma_y \rho^* \sigma_y \bigotimes \sigma_y \tag{3.8}$$

where $*$ means complex conjugate. When $C=0$, it means that the two atoms are not entangled; when $C = 1$, it means that the quantum atom is in the maximum entangled state.

According to the reduced density matrix of atoms Ω_1, Ω_2 and the definition of concurrency, we can know that the simplified expression of concurrency is

$$\begin{aligned} C_1(t) &= 2|\rho_{23}| - 2\sqrt{\rho_{11}\rho_{44}} \\ &= |\ sin(2\theta)\ cos(gt)\left[cos^2\phi cos\left(\sqrt{2}gt\right) + sin^2\phi\right] \\ &- \sqrt{sin^2 2\theta\ cos^2\phi\ sin^2(gt)\left[cos^2\left(\sqrt{2}gt\right) + sin^2\phi\ sin^2(gt)\right]} \end{aligned} \tag{3.9}$$

The linear entropy of the two atoms Ω_1, Ω_2

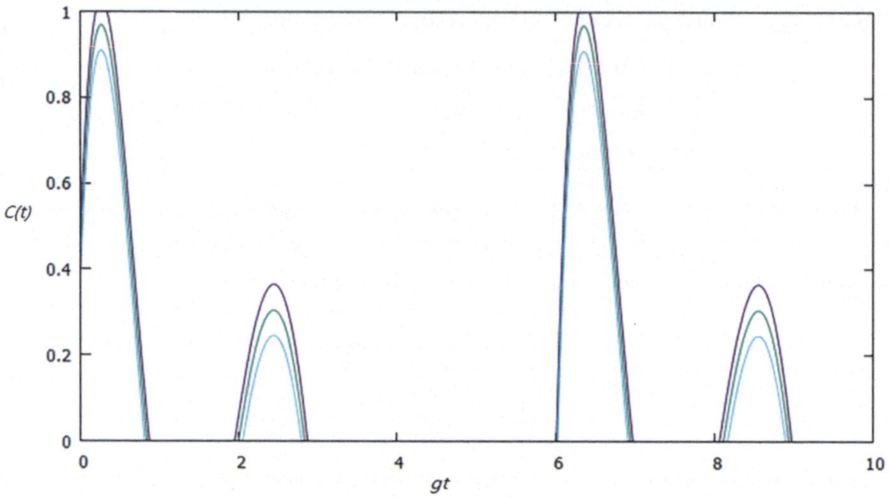

Fig. 3.2 Evolution of concurrency over time 1

$$\begin{aligned} S(t) &= 1 - T_r\left[\rho^2 \Omega_2 \Omega_1(t)\right] \\ &= 1 - \left[\rho_{11}^2(t) + \rho_{22}^2(t) + \rho_{44}^2(t) + 2\rho_{23}(t)\rho_{32}(t)\right]. \end{aligned} \tag{3.10}$$

Therefore, when $S = 0$, it is unentangled, and when $S = 0.75$, it is the maximum entangled state.

The evolution characteristics of the entanglement concurrency between two atoms Ω_1, Ω_2 over time are shown in Fig. 3.2.

As can be seen from Fig. 3.2, when cavity field system is in the maximum entangled state $\phi = \pi/4$, the entanglement evolution characteristics of atoms are characterized by different entangled states. The phenomenon called "entanglement death" occurs when the entanglement between two atoms suddenly decreases to zero and remains there for some time before the entanglement recovers. The main reason is that the atoms in the two-mode cavity interact with the cavity mode field, so that the information energy between the atoms and the cavity is constantly transferred, resulting in the entanglement between the two atoms transferred to the two-mode cavity. According to Fig. 3.2, the time of entanglement death has little relation with the initial entanglement degree between atoms, and the smaller the initial entanglement degree is, the longer the entanglement death will last.

When the atom is initially in the maximum entangled state $\theta = \pi/4$ and the cavity field parameters ϕ are set to different values, the evolution characteristics of concurrency over time are shown in Fig. 3.3.

As can be seen from Fig. 3.3, entanglement death still occurs, but the duration of entanglement death is related to the degree of entanglement in the cavity field. The smaller the initial degree of entanglement of atoms, the shorter the time for the first entanglement death of two atoms.

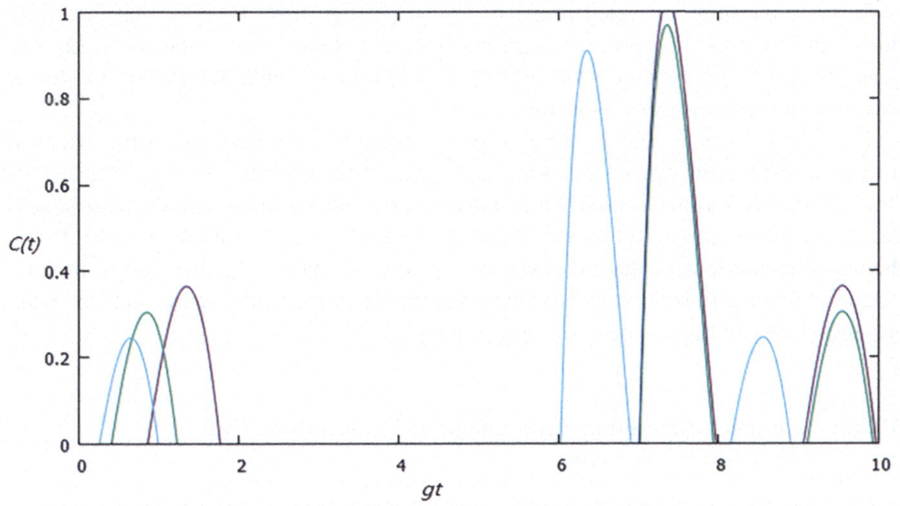

Fig. 3.3 Evolution of concurrency over time 2

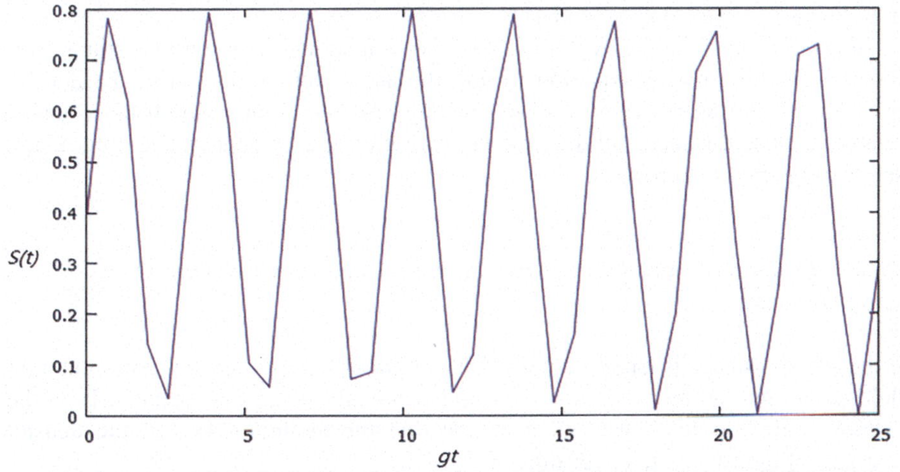

Fig. 3.4 Evolution of linear entropy of two atoms over time

When the atom is initially in the maximum entangled state $\theta = \pi/4$ and the cavity field is also in the maximum entangled state $\phi = \pi/4$, the linear entropy of the two atoms Ω_1, Ω_2 evolves over time as shown in Fig. 3.4:

The entanglement between two atoms Ω_1, Ω_2 and the cavity field f is reflected by the linear entropy. As can be seen from Fig. 3.4, when atoms and cavity field interact, energy exchange will occur between them, resulting in entanglement. When the interaction progresses to a certain extent, linear entropy will reach the maximum value $S = 0.75$. At this time, the two atoms Ω_1, Ω_2 and the cavity field f appear to have the maximum entangled property. It can be seen from Fig. 3.4 that the entire linear entropy evolution process is greater than zero, that is to say, the atoms Ω_1, Ω_2, subsystem, and cavity field f subsystem always remain entangled during the linear entropy evolution process with time.

Using concurrency and linear entropy as measures, we investigate the effect of initial entanglement degree of atom and cavity field on concurrency. The results show that the duration of entanglement death is not related to the initial entanglement degree between atoms, but to the initial entanglement degree of cavity field. When the initial entanglement degree of the cavity field is smaller, the duration of the first entanglement death is shorter, but the two atoms Ω_1, Ω_2 and the cavity field f remain entangled during the whole evolution process.

3.1.1.2 Characterization of Entanglement Evolution of Two Three-Level Systems

We assume that there are two three-level atoms A, B and then put them in a vacuum chamber, and every atom will have excited states $|e\rangle$ and $|a\rangle$ and the ground state $|g\rangle$. There can be galvanic transitions between atoms that can emit spontaneous radiation, i.e., $|e\rangle \rightarrow |g\rangle$, $|a\rangle \rightarrow |g\rangle$ [5].

In order to analyze the evolution law of entanglement between two atoms over time, we use the master equation to find the analytical expression of the density operator of two atoms over time by placing two atoms at a far distance and a relatively close distance and describe the evolution law of entanglement through a series of operational changes.

Characterization of Entanglement Evolution of Three-Level Systems Over Long Distances

If the space distance is larger than the wavelength of the spontaneous emission light field of the two atoms, and each atom independently decays spontaneously in the vacuum radiation field, we can conclude that the evolution law of the density operator of the two atoms over time is

$$
\begin{aligned}
&\rho_{22}(t)=\rho_{22}(0)e^{-(x_1+x_2)t}\\
&\rho_{33}(t)=[\rho_{11}(0)(1-e^{-x_1t})+\rho_{22}(0)(1-e^{-x_2t})+\rho_{33}(0)e^{-x_1t}\\
&\rho_{44}(t)=\rho_{44}(0)e^{-(x_1+x_2)t}\\
&\rho_{66}(t)=[\rho_{55}(0)(1-e^{-x_2t})+\rho_{44}(0)(1-e^{-x_1t})+\rho_{66}(0)e^{-x_2t}\\
&\rho_{88}(t)=[\rho_{55}(0)(1-e^{-x_2t})+\rho_{22}(0)(1-e^{-x_1t})+\rho_{88}(0)e^{-x_2t}\\
&\rho_{15}(t)=\rho_{15}(0)e^{-(x_1+x_2)t}\ \rho_{19}(t)=\rho_{19}(0)e^{-x_1t}\ \rho_{59}(t)=\rho_{59}(0)e^{-x_2t}
\end{aligned}
\tag{3.11}
$$

where x is the spontaneous emission rate. We use the density matrix method to perform partial transpose operation on the two atoms obtained at any time, and the eigenvalues we obtained are as follows:

$$E_1 = \rho_{11}(t)\ E_2 = \rho_{55}(t)\ E_3 = \rho_{99}(t)$$
$$E_4 = 1/2\left[(\rho_{22}(t) + \rho_{44}(t)) - \sqrt{(\rho_{22}(t) + \rho_{44}(t))^2 + 4\left(|\rho_{15}(t)|^2 - \rho_{22}(t)\rho_{44}(t)\right)}\right]$$
$$E_5 = 1/2\left[(\rho_{22}(t) + \rho_{44}(t)) + \sqrt{(\rho_{22}(t) + \rho_{44}(t))^2 + 4\left(|\rho_{15}(t)|^2 - \rho_{22}(t)\rho_{44}(t)\right)}\right]$$
$$E_6 = 1/2\left[(\rho_{33}(t) + \rho_{77}(t)) - \sqrt{(\rho_{33}(t) + \rho_{77}(t))^2 + 4\left(|\rho_{19}(t)|^2 - \rho_{33}(t)\rho_{77}(t)\right)}\right].$$
$$E_7 = 1/2\left[(\rho_{33}(t) + \rho_{77}(t)) + \sqrt{(\rho_{33}(t) + \rho_{77}(t))^2 + 4\left(|\rho_{19}(t)|^2 - \rho_{33}(t)\rho_{77}(t)\right)}\right]$$
$$E_8 = 1/2\left[(\rho_{66}(t) + \rho_{88}(t)) - \sqrt{(\rho_{66}(t) + \rho_{88}(t))^2 + 4\left(|\rho_{59}(t)|^2 - \rho_{66}(t)\rho_{88}(t)\right)}\right]$$
$$E_9 = 1/2\left[(\rho_{66}(t) + \rho_{88}(t)) + \sqrt{(\rho_{66}(t) + \rho_{88}(t))^2 + 4\left(|\rho_{59}(t)|^2 - \rho_{66}(t)\rho_{88}(t)\right)}\right] \tag{3.12}$$

After analyzing, we can know that the eigenvalues of E_4, E_6, and E_8 in the above formula may be negative, which can be expressed as

$$N(t) = m\{-E_4, 0\} + m\{-E_6, 0\} + m\{-E_8, 0\}. \tag{3.13}$$

When one of $|\rho_{15}(t)|^2 > \rho_{22}(t)\rho_{44}(t)$, $|\rho_{19}(t)|^2 > \rho_{33}(t)\rho_{77}(t)$, and $|\rho_{59}(t)|^2 > \rho_{66}(t)\rho_{88}(t)$ is satisfied, there is an entanglement between the two atoms.

When a moment exists, $|\rho_{15}(t)|^2 = \rho_{22}(t)\rho_{44}(t)$, $|\rho_{19}(t)|^2 = \rho_{33}(t)\rho_{77}(t)$, or $|\rho_{59}(t)|^2 = \rho_{66}(t)\rho_{88}(t)$, the entanglement suddenly changes, resulting in a "sudden death" or "recovery" of entanglement depending on how the two-atom density operator changes over time.

Since the entanglement evolution is related to the initial state of the cavity field, the nine-dimensional identity matrix composed of the identity matrix I and the state function $|\Phi\rangle$ is selected as an example:

$$\begin{gathered}|\Phi\rangle = \cos\theta\sin\phi|1\rangle + \sin\theta\sin\phi|5\rangle + \cos\phi|9\rangle \\ \rho_0 = [(1-P)/9]I + P|\phi\rangle\langle\phi|\end{gathered} \tag{3.14}$$

In the above formula, θ and ϕ take different values. When $\theta = 0$ and $\phi = \pi/4$, $|\Phi\rangle = \sqrt{2}/2(|1\rangle + |9\rangle)$, it means the initial state of the system consists of $|1\rangle$, $|9\rangle$ and identity matrix I. What we can know from the formula of eigenvalue is that the character of system entanglement can be characterized by eigenvalue E_6. The evolution of entanglement over time is shown in Fig. 3.5:

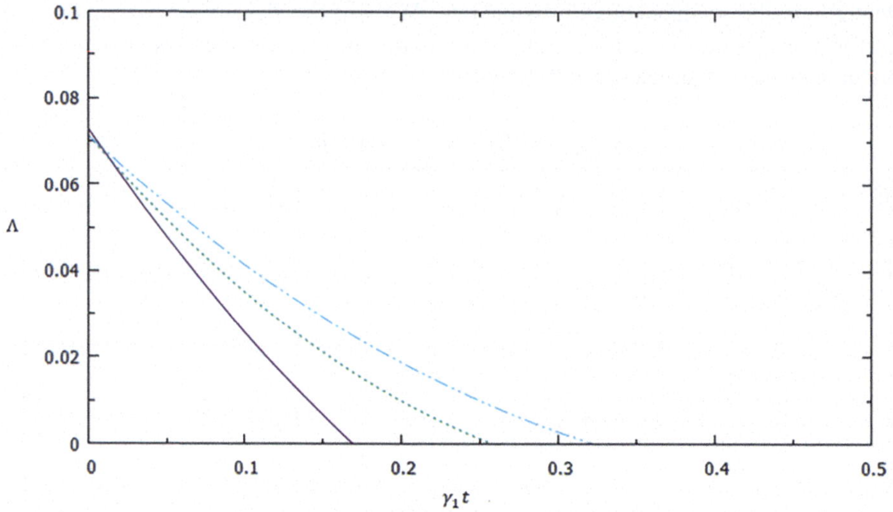

Fig. 3.5 Evolution of entanglement over time

As can be seen from Fig. 3.5, entanglement death is caused by the reduction of spontaneous radiation between atoms, and the time of entanglement death is related to the rate of atomic decay. In other words, the higher the decay rate, the earlier the entanglement death time, which indicates that the decay rate of the excited state will directly affect the entanglement degradation rate.

When $\theta = \pi/2$ and $\phi = \pi/4$, $|\Phi\rangle = \sqrt{2}/2(|5\rangle + |9\rangle)$, it means the initial state of the system consists of $|5\rangle$, $|9\rangle$ and identity matrix I. At this time, eigenvalue $E_4 > 0$ and $E_6 > 0$. According to the formula of eigenvalue, the characteristic of system entanglement can be described by eigenvalue E_8. Entanglement death also occurs at a certain time, and the time of entanglement death is related to the rate of atomic decay. The higher the decay rate, the earlier the entanglement death time. Thus, the rate of entanglement degradation is related to the decay rate of the atomic excited state.

When $\theta = \pi/4$ and $\phi = \pi/2$, $|\Phi\rangle = \sqrt{2}/2(|1\rangle + |5\rangle)$, it means the initial state of the system consists of $|1\rangle$, $|5\rangle$ and identity matrix I. According to the formula of eigenvalue, the characteristic of system entanglement can be described by eigenvalue E_4. At this time, $\rho_{22}(t)\rho_{44}(t)$ and $|\rho_{15}(t)|^2$ have same decay rate. If $|\rho_{15}(t)|^2 > \rho_{22}(t)\rho_{44}(t)$, then $E_4 < 0$, showing that entanglement always exists in the evolution of the system. If $|\rho_{15}(t)|^2 < \rho_{22}(t)\rho_{44}(t)$, the initial state is unentangled and there is no entanglement between the two atoms.

When $\theta = \pi/4$ and $\phi = \pi/4$, $|\Phi\rangle = 1/2(|1\rangle + |5\rangle) + (\sqrt{2}/2)\,|9\rangle$, it means the initial state of the system consists of $|1\rangle$, $|5\rangle$, $|9\rangle$ and identity matrix I.

According to the above analysis, if $E_4 > 0$, then entanglement is always zero. If at a certain moment, $E_6 = 0$ and $E_8 = 0$, the evolution characteristics of entanglement

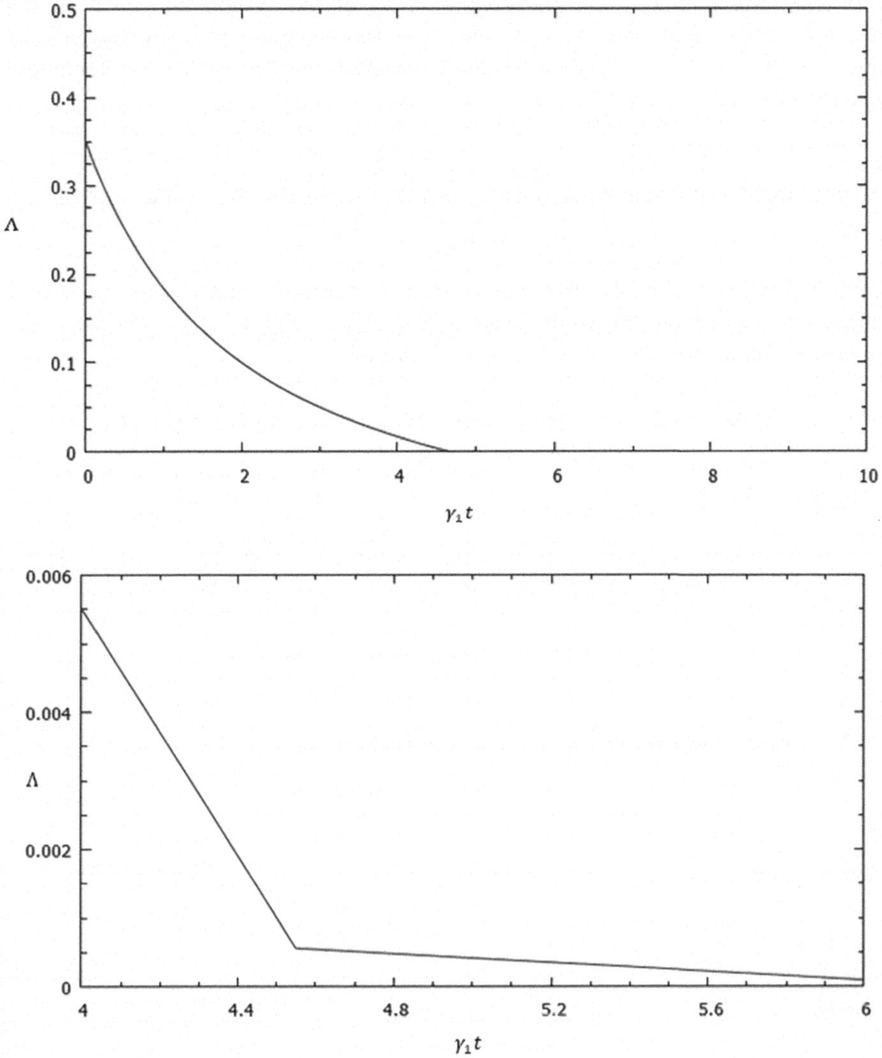

Fig. 3.6 Evolution of quantum properties over a long distance with time

death caused by entanglement are shown in Fig. 3.6. Two diagrams are used to represent the entanglement evolution under two parameters.

When eigenvalue $E_6 = 0$ and $E_8 = 0$, entanglement death occurs. If $E_4 < 0$, the entanglement characterized with E_4 will always exist. At this moment, if certain times E_f and E_s both are zero, the entanglement will have mutation, but the system only undergoes a momentary mutation, and the entanglement death phenomenon will not occur for the existence of E_4.

When eigenvalue $E_6 = 0$, the entanglement of the system will have its first mutation. When eigenvalue $E_6 = 0$ and $E_8 = 0$, entanglement of the system will have a second mutation and then gradually weaken, but there will be no entanglement death.

Characterization of Entanglement Evolution of Three-Level System at Close Range

In this system, we select the mix state of identity matrix I, excited state $|5\rangle$, and the coherent superposition state of the ground state $|5\rangle$ as the initial state; then we obtain the law of characterization of entanglement evolution:

$$\begin{gathered}\rho_{11}(t)=1/9e^{-2xt}(1-P)=\rho_{22}(t)=\rho_{44}(t)\ \rho_{55}(t)=1/18e^{-2xt}(2+7P)\\ \rho_{33}(t)=1/36e^{-2xt}(1-P)\left(1+3e^{-2xt}+6xt\right)=\rho_{77}(t)\\ \rho_{66}(t)=1/36e^{-2xt}(1-P)\left(1+3e^{-2xt}+6xt\right)+P/2xte^{-2xt}=\rho_{88}(t)\\ \rho_{37}(t)=\rho_{73}(t)=1/12e^{-2xt}(1-P)\left(1-e^{2xt}+2xt\right)\\ \rho_{86}(t)=\rho_{68}(t)=1/12e^{-2xt}(1-P)\left(1-e^{2xt}+2xt\right)+P/2xte^{-2xt}\\ \rho_{59}(t)=\rho_{95}(t)=1/2e^{-xt}P\ \rho_{99}(t)=1-\sum_{i=1}^{8}\rho_{ii}(t)\end{gathered} \tag{3.15}$$

The evolution of entanglement over time is shown in Fig. 3.7:

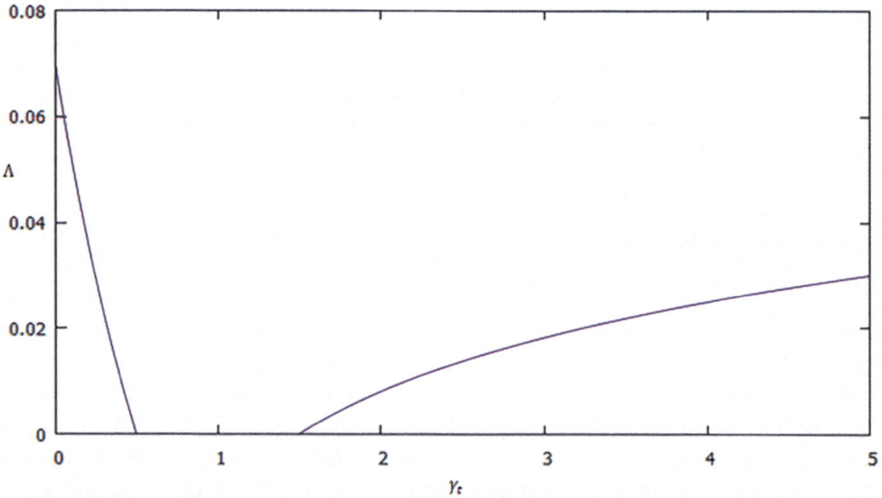

Fig. 3.7 Evolution of entanglement over time

As can be seen from Fig. 3.7, the initial entanglement can decay by spontaneous radiation which will cause sudden death to occur at any given moment. However, due to the cooperation effect between atoms of vacuum radiation, after a period of entangled death, the system will slowly show signs of recovery and gradually increase and finally become a constant state. The initial entanglement is characterized by eigenvalues E_8:

$$E_8 = \frac{1}{2}\left[\rho_{66}(t) + \rho_{88}(t) - \sqrt{(\rho_{66}(t) + \rho_{88}(t))^2 + 4(|\rho_{59}(t)|^2 - |\rho_{66}(t)|\rho_{88}(t))}\right]. \quad (3.16)$$

According to the change rule of density operator over time, $|\rho_{59}(t)|^2 > \rho_{66}(t)\rho_{88}(t)$ and eigenvalue $E_8 < 0$ suggest an entanglement between the two atoms. When $|\rho_{59}(t)|^2 = \rho_{66}(t)\rho_{88}(t)$, $E_8 = 0$, the phenomenon of entanglement disappearance will occur which is caused by entanglement sudden death. Therefore, when the number of particles of the transition state is less than the number of particles of the coherent state, the system will have entanglement but the entanglement death will not occur. When the number of particles in the transition state is equal to the number of particles in the coherent state, the system will exhibit entanglement mutation and death phenomenon.

In order to solve the problem of entanglement death, we used E_3 to describe the recovery process of entanglement. When the value of E_3 gradually increases until stable, the entanglement analysis after stable is as follows:

$$\begin{aligned} &\rho_{33}(\infty) = \rho_{77}(\infty) = \rho_{66}(\infty) = \rho_{88}(\infty) = -\rho_{37}(\infty) = -\rho_{68}(\infty) = \frac{P-1}{12} \\ &\rho_{11}(\infty) = \rho_{22}(\infty) = \rho_{44}(\infty) = \rho_{55}(\infty) = \rho_{59}(\infty) = 0\ \rho_{99}(\infty) = \frac{2+P}{3} \end{aligned} \quad (3.17)$$

Thus, we get eigenvalues E_3 as follows:

$$E_3 = \frac{1}{2}\left[\rho_{99}(\infty) - \sqrt{\rho_{99}^2(\infty) + 4\left(|\rho_{37}(\infty)|^2 + |\rho_{68}(\infty)|^2\right)}\right] \quad (3.18)$$

All in all, the single particle of transition state is the core and decisive factor of entanglement steady state. As long as the number of single particles in transition state is not zero, entanglement will always exist. The entanglement before death is determined by the factors of the coherent superposition state of two atoms and the number of single particles in the transition state, while the entanglement after recovery is determined by the number of particles in the single particle transition state. We should consider above factors when designing the system.

3.1.2 Characterization of Quantum Decoherence

Because the key of building immune noise model is to solve the quantum state decoherence caused by coupling between mixed physical system and environment, the first problem to be solved is how to find and construct coherent maintained states in noise channels based on multi-degree of freedom (such as time, space, and frequency) characteristics, so that the obtained quantum state is always in the maximum entangled state and the high standard service for the quantum teleportation process is always maintained.

To "observe" how a quantum moves, we must have macroscopic objects interacting with it to form entanglement, which inevitably leads to quantum decoherence, a phenomenon that has a large effect on quantum state transfer. We clarify the basis of entanglement death problem and describe the quantum decoherence characteristics between atom and cavity field in T-C model and between two cavity fields and two two-level atoms in Jaynes-Cummings (J-C) model, removing obstacles for the establishment of quantum entanglement evolution model.

3.1.2.1 Characterization of Quantum Decoherence of T-C Model

The Tavis-Cummings (T-C) model [6] is a model in which two atoms interact with the cavity field. We assume that the atom uses two two-level atoms and the cavity field uses a single-mode light field. If the distance between the atoms is shorter than the wavelength of the light field, then the Hamilton quantity of the system is $\hbar = 1$.

$$H = wa^{+}a + \frac{1}{2}\sum_{i=1}^{2} w_i\sigma_i^{z} + \sum_{i=1}^{2} g_i\left(a_i^{+}\sigma_i^{-} + a_i\sigma_i^{+}\right) + \Omega \sum_{i=i=1}^{2} \sigma_i^{-}\sigma_j^{+} \tag{3.19}$$

where w_1 and w_2 are the intrinsic transition frequencies of the two atoms, a_i is the annihilation operator of the light field, a_i^{+} is the generation operator of the light field, $\sigma_i^{z}, \sigma_i^{\pm}$ are the spin operator of the atom, $|e_i\rangle$ is the excited state of the atom, $|g_i\rangle$ is the ground state of the atom, and Ω represents the interaction between atoms.

It is assumed that in the initial state $w_1 = w_2 = w$, $g_1 = g_2 = g$, the two atoms are in the state of entanglement and the light field also has the state of particle number. Then the system state is

$$|\Phi(0)\rangle = (\cos\theta|e_1, g_2\rangle + \sin\theta|g_1, e_2\rangle) \bigotimes |n\rangle. \tag{3.20}$$

When $t > 0$, the evolution of the system state is

$$\begin{aligned}|\Phi(t)\rangle = C_1(t)|e_1,e_2,n-1\rangle + C_2(t)|e_1,e_2,n\rangle \\ +C_3(t)|g_1,e_2,n\rangle + C_4(t)|g_1,g_2,n+1\rangle\end{aligned}. \tag{3.21}$$

Plugging in the Schrodinger equation $i\hbar\frac{\partial|\Phi(t)\rangle}{\partial t} = H|\Phi(t)\rangle$, we get

$$\begin{aligned}C_1(t) &= \cos\theta + \sin\theta)g\sqrt{n}\left(e^{iat} - e^{ibt}\right)/\Delta \\ C_2(t) &= \left[-a(\cos\theta + \sin\theta)e^{iat} + b(\cos\theta + \sin\theta)e^{ibt}//2\Delta\right. \\ &\quad +\left[(\cos\theta - \sin\theta)/2\right|e^{i\Omega t} \\ C_3(t) &= \frac{\left[-a(\cos\theta + \sin\theta)e^{iat} + b(\cos\theta + \sin\theta)e^{ibt}\right]}{2\Delta} \\ &\quad +\left[(\sin\theta - \cos\theta)/2\right]e^{i\Omega t} \\ C_4(t) &= (\cos\theta + \sin\theta)g\sqrt{n+1}\left(e^{iat} - e^{ibt}\right)/\Delta\end{aligned}. \tag{3.22}$$

The reduced density matrix of the two atoms is

$$\rho_{12}(t) = \begin{bmatrix}|C_4(t)|^2 & 0 & 0 & 0 \\ 0 & |C_3(t)|^2 & C_3(t)C_2(t) & 0 \\ 0 & C_2(t)C_3(t) & |C_2(t)|^2 & \\ 0 & 0 & 0 & |C_1(t)|^2\end{bmatrix}. \tag{3.23}$$

After using linear entropy for measurement, we define linear entropy as follows:

$$S = 1 - T_r\left\{\rho_{1,2}^2(t)\right\}. \tag{3.24}$$

According to the above formula analysis, when $S = 0$, the two subsystems are in the maximum entangled state. When $S = 1$, the two subsystems are separated from each other. When the value of S is between 0 and 1, the two subsystems are in the entangled state.

When the initial states of two atoms change, we analyze the law of entanglement evolution and decoherence between the atoms and the cavity field which is the change law of entanglement quantity S with θ.

When $g = 1, \Omega = 1, n = 0$, and $\theta = 0$, there is an entanglement evolution between atoms and the field as shown in Fig. 3.8:

When $g = 1$, $\Omega = 1$, $n = 0$, and $\theta = \pi/4$, there is an entanglement evolution between atoms and the field as shown in Fig. 3.9:

When $g = 1$, $\Omega = 1$, $n = 0$, and $\theta = 3\pi/4$, there is an entanglement evolution between atoms and the field as shown in Fig. 3.10:

When $g = 1$, $\Omega = 1$, $n = n$, and $\theta = 7\pi/8$, there is an entanglement evolution between atoms and the field as shown in Fig. 3.11:

Figures 3.8, 3.9, 3.10 and 3.11 show that the degree of entanglement changes periodically with time. It increases from zero, but cannot reach 1 at the maximum,

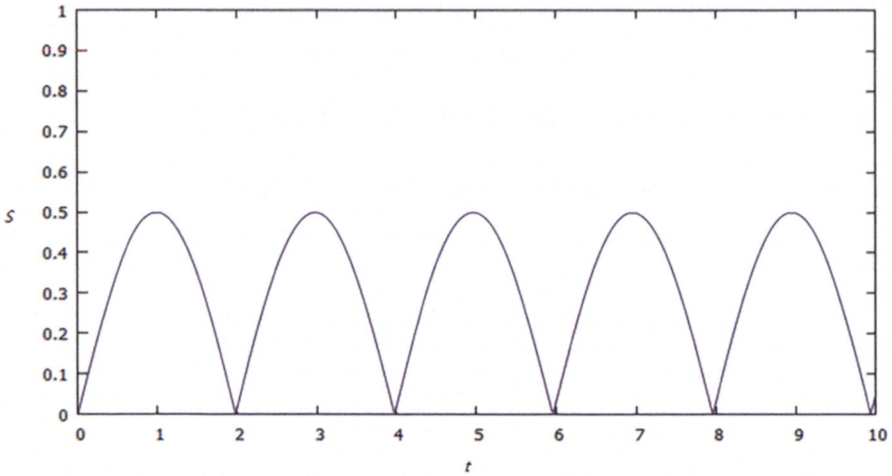

Fig. 3.8 Entanglement evolution between atoms and the field when $\theta = 0$

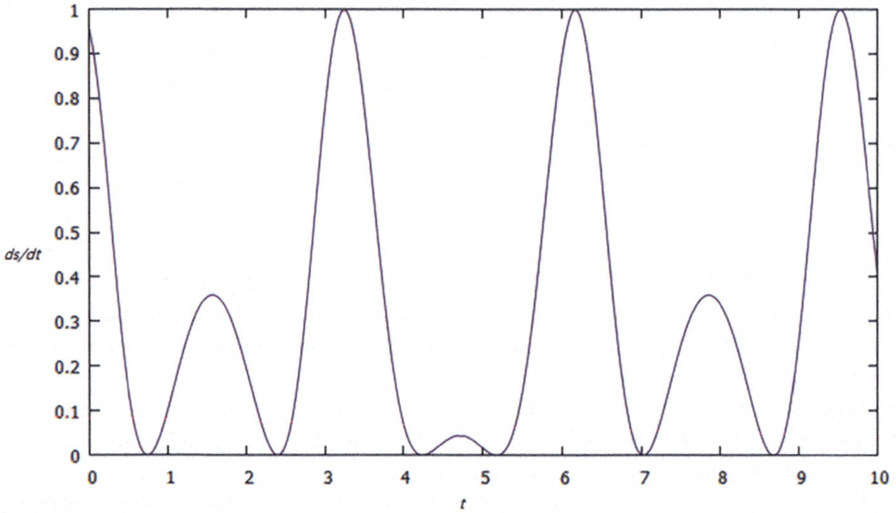

Fig. 3.9 Entanglement evolution between atom and field when $\theta = \pi/4$

and then decreases gradually until it reaches zero, then increases again, and then decreases again.

According to Fig. 3.11, when the degree of entanglement is zero, $\cos\theta = \sin\theta = -\frac{\sqrt{2}}{2}$ which means the initial state is equal to the eigenstate of the Hamilton quantity.

The probability of the radiated atom and the excited atom is equal, and the two atoms will have interference cancellation. It can be seen that the two atoms are

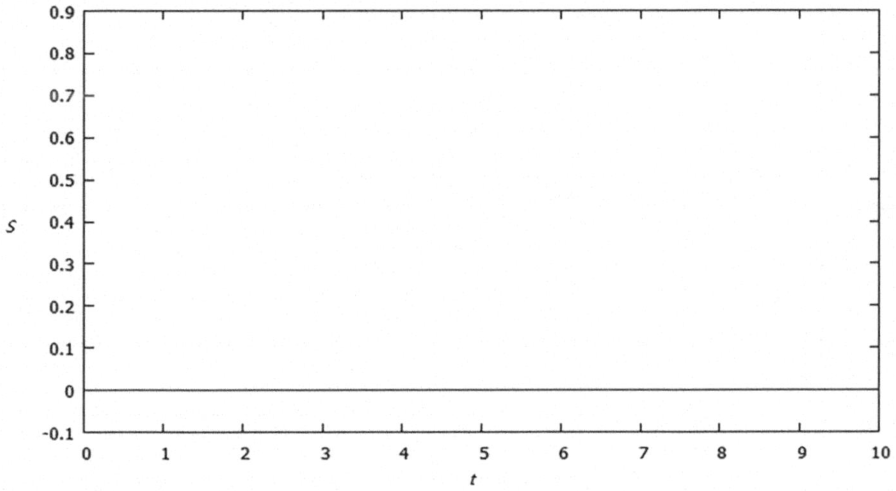

Fig. 3.10 Entanglement evolution between atoms and the field when $\theta = 3\pi/4$

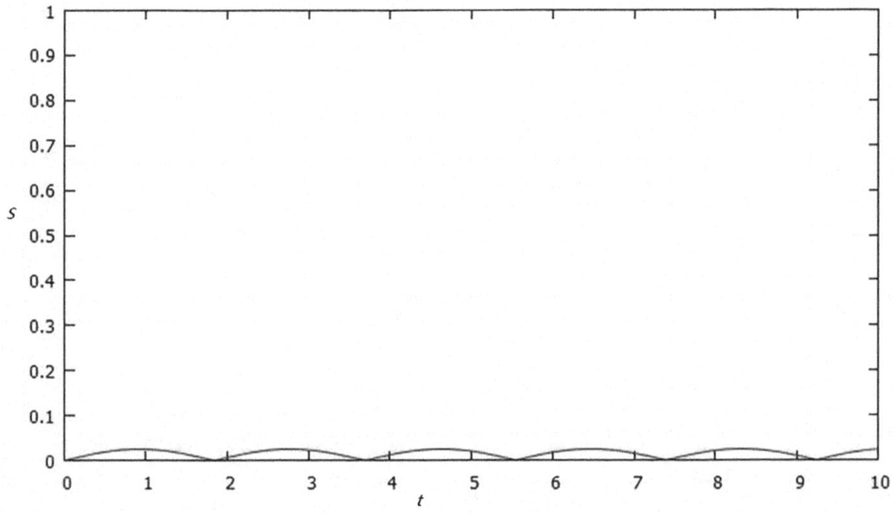

Fig. 3.11 Entanglement evolution between atoms and the field when $\theta = 7\pi/8$

always in the maximum entanglement state. In other words, the quantum state remains in a separate state after interaction, so the entanglement degree is zero.

The coherence of a quantum system is represented by the off-diagonal terms of the density matrix to reflect this coherence characterization process. When the decoherence effect occurs, different evolution characteristics will be presented. At this time, the off-diagonal term of the density matrix becomes zero. Therefore, the

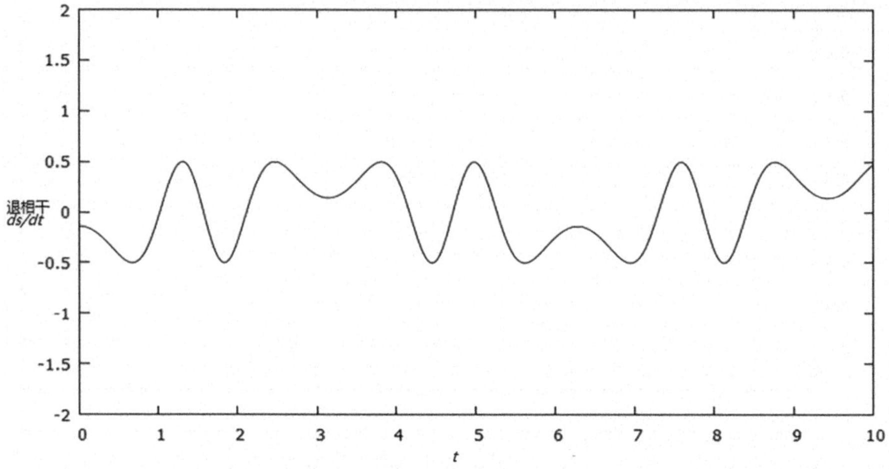

Fig. 3.12 Decoherence curve with time when $\theta = \pi/6$, $n = 1$

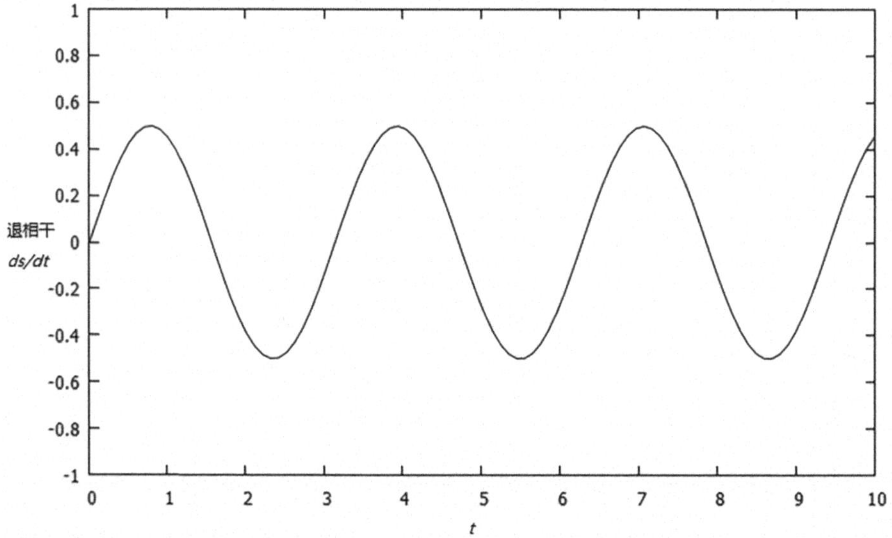

Fig. 3.13 Decoherence curve with time when $\theta = 7\pi/8$, $n = 0$

time scale of decoherence can well measure the evolution speed of the off-diagonal term of the density matrix over time. When $g = 1$, $\Omega = 1$, considering the variation of the angle θ and photon number, the time-dependent decoherence curve is shown in Fig. 3.12. This process is the characterization of decoherence.

It can be seen from Figs. 3.12 and 3.13 that when $n = 1$ the amplitude of entanglement degree was relatively large and the entanglement period was short, while the amplitude of entanglement degree was relatively small but the entanglement period was long when $n = 0$. Therefore, if a larger degree of entanglement is required, a bigger value of n should be selected, and vice versa.

Thus, our analysis shows that the state of the light field has a great influence on the decoherence period and change of the entanglement between the atom and the field.

3.1.2.2 Characterization of Quantum Decoherence of J-C Model

We assume that there are two entangled cavity fields C_1 and C_2 and two two-level atoms A and B. Place the atom A in the cavity field C_1 and place the atom B in the cavity field C_2. According to the J-C model theory, the Hamilton quantity of the system under the approximate rotation wave can be expressed as

$$H = g\left(a_1\sigma_+ + a_1^+\sigma_- + a_2\sigma_+ + a_2^+\sigma_-\right) \tag{3.25}$$

where a^+ is the generation operator of the cavity field, a is the annihilation operator of the cavity field, g is the coupling strength coefficient, and $a^{\pm}$ is the spin operator; then the evolution equation of the system is

$$\begin{aligned}\rho_1 = {} & -i[H,\rho] = \\ & -ig\left(a_1\sigma_+\rho + a_1^+\sigma_-\rho - \rho a_1\sigma_+ - \rho a_1\sigma_- + a_2\sigma_+\rho + a_2^+\sigma_-\rho - \rho a_2\sigma_+ - \rho a_2\sigma_-\right)\end{aligned} \tag{3.26}$$

$$\rho_2 = -K_1\left(a_1^+a_1\rho - 2a_1\rho a_1^+ + \rho a_1^+a_1\right) - K_2\left(a_2^+a_2\rho - 2a_2\rho a_2^+ + \rho a_2^+a_2\right) \tag{3.27}$$

where $k_{1,2}$ is the dissipation factor. The wave function of the initial state of the system when $t = n$ is

$$|\phi\rangle_{C_1C_2AB} = \frac{1}{\sqrt{2}}(|01\rangle + |10\rangle)_{12} \bigotimes |gg\rangle_{12}. \tag{3.28}$$

Thus, we get the density matrix operator of the system as follows:

$$\begin{aligned}|\Psi\rangle = {} & \Omega_1|0_11_2g_1g_2\rangle\langle 0_11_2g_1g_2| + \Omega_2|0_10_2g_1e_2\rangle\langle 0_10_2g_1e_2| + \Omega_3|1_10_2g_1g_2\rangle\langle 1_10_2g_1g_2| \\ & + \Omega_4|0_10_2e_1g_2\rangle\langle 0_10_2e_1g_2| + \Omega_5|0_11_2g_1g_2\rangle\langle 1_10_2g_1g_2| + \Omega_5|1_10_2g_1g_2\rangle\langle 0_11_2g_1g_2| \\ & + \Omega_6|0_10_2g_1e_2\rangle\langle 0_10_2e_1g_2|\end{aligned} \tag{3.29}$$

In Formula (3.29),

$$\begin{aligned}\Omega_1 &= \left(1-\frac{e^{-k_1 t}}{2}\right)e^{-k_2 t}\cos^2 gt, \Omega_2 = \left(\sin^2 gt\right)e^{-k_2 t}\left(1-\frac{e^{-k_1 t}}{2}\right)\\ \Omega_3 &= \left(\cos^2 gt\right)e^{-k_1 t}\left(1-\frac{e^{-k_2 t}}{2}\right), \Omega_4 = \left(\sin^2 gt\right)e^{-k_1 t}\left(1-\frac{e^{-k_2 t}}{2}\right)\\ \Omega_5 &= \left[(\cos gt)e^{-k_1\frac{t}{2}}e^{-k_2\frac{t}{2}}\right]/2\\ \Omega_6 &= \frac{1}{2}\left(e^{-k_1\frac{t}{2}}\sin t - \frac{k_1 e^{-k_1\frac{t}{2}}}{2g} + k_1/2g\right)\end{aligned} \tag{3.30}$$

Combining the above formulas, we obtain two atomic reduced density matrices:

$$\begin{aligned}\rho(t)_{AB} = T_{rC_1C_2}\rho(t) &= (\Omega_1+\Omega_3)|g_1g_2\rangle\langle g_1g_2| + \Omega_2|g_1e_2\rangle\langle g_1e_2|\\ &+\Omega_4|e_1g_2\rangle\langle e_1g_2| + \Omega_6|g_1e_2\rangle\langle e_1g_2| + \Omega_6|e_1g_2\rangle\langle g_1e_2|\end{aligned} \tag{3.31}$$

The decoherence of quantum system is measured by the off-diagonal element of density matrix. When the off-diagonal element is not zero, the decoherence phenomenon exists. The decoherence factors of two atoms are obtained by analysis, as shown below:

$$\begin{aligned}F(g,k,t) &= \frac{1}{2}\left(e^{-k_1\frac{t}{2}}\sin gt - \left(K_1 e^{-k_1\frac{t}{2}}\right)/2g + k_1/2g\right)\\ &\left(e^{-k_1\frac{t}{2}}\sin t - \left(K_2 e^{-k_2\frac{t}{2}}\right)/2g + k_2/2g\right)\end{aligned} \tag{3.32}$$

According to the above equation, the decoherence of the quantum system is related to the dissipation factor and coupling constant of the cavity. That can also be shown in Fig. 3.14.

When the dissipation factor $k_1 = k_2 = 0g$, the decoherence factor between two atoms changes with gt shown in Fig. 3.14. According to the analysis from Fig. 3.14, when there is no dissipation factor, the amplitude of the decoherence factor of the quantum system does not change with time, with a range of 0–1, and the period is the same.

When the dissipation factor $k_1 = k_2 = 0.1g$, the decoherence factor between two atoms changes with gt shown in Fig. 3.15.

When the dissipation factor $k_1 = k_2 = 0.2g$, the decoherence factor between two atoms changes with gt shown in Fig. 3.16.

According to the comparative analysis from Figs. 3.15 and 3.16, in the presence of dissipative system, the decoherence factor between the two atoms shows amplitude oscillation with reduced amplitude, and after a period of amplitude oscillation, the amplitude decreases to zero, and complete decoherence phenomenon appears indicating that the dissipative system affects the decoherence of the quantum system.

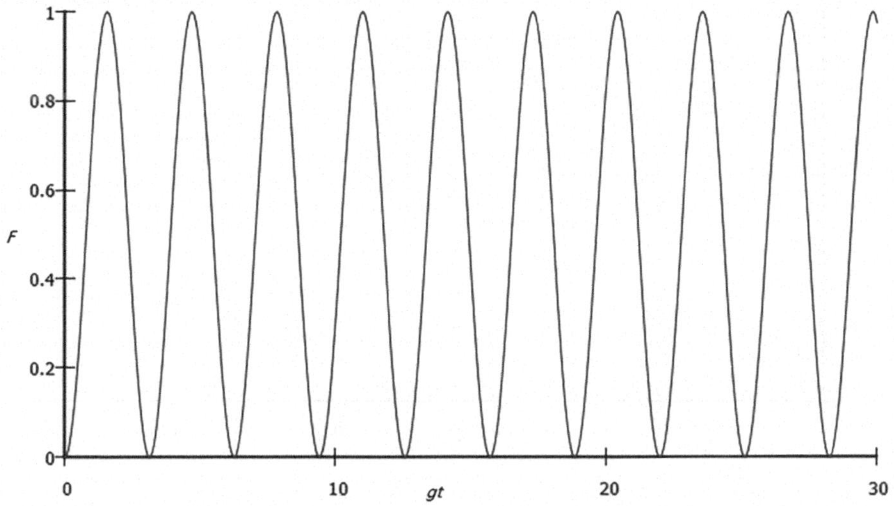

Fig. 3.14 Curve of the decoherence factor between two atoms changes with *gt*

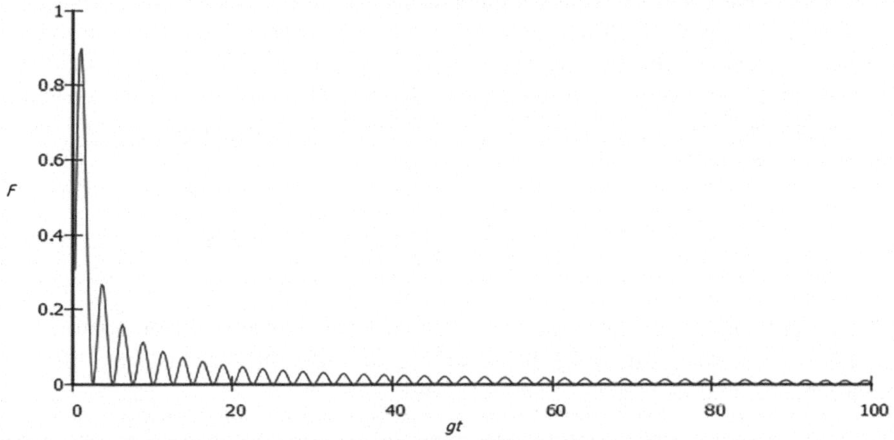

Fig. 3.15 Curve of the decoherence factor between two atoms changes with *gt*

3.1.3 Evolution of Quantum Entanglement in Local Common Mode

Based on the characterization model of entanglement death and quantum decoherence in building open system, analyzing the effects of different noises on the entanglement evolution characteristics of entangled mixed states in local common mode especially when in local common noise and how to characterize the

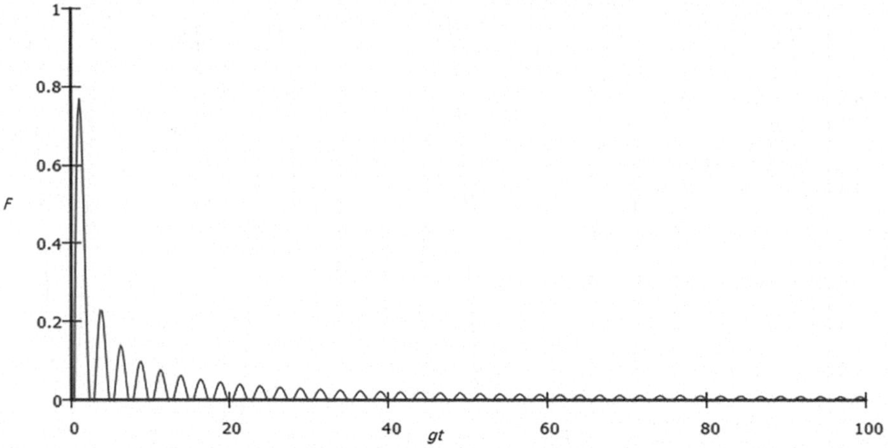

Fig. 3.16 Curve of the decoherence factor between two atoms changes with *gt*

process of entanglement sudden death and resurrection is the core of the construction of quantum entanglement evolution model.

We use the unitary and non-unitary evolution theories of the quantum master equation to construct the density matrix of the open quantum system, calculate the corresponding degree of concurrency, and describe the evolution of the quantum system affected by locally independent and locally common noises. The master equation is as follows:

$$\frac{d\rho(t)}{dt} = \sum_{\mu} \left(L_{\mu}\rho(t)L_{\mu}^{\dagger} - \frac{1}{2}L_{\mu}^{\dagger}L_{\mu}\rho(t) - \frac{1}{2}\rho(t)L_{\mu}^{\dagger}L_{\mu} \right) \tag{3.33}$$

Different noise generation operators L_{μ} are selected to reflect the dissipative part of the system state caused by the coupling of different systems and different environments, and its properties are as follows:

If energy dissipation is caused by the coupling between the system and the zero-temperature environment, then $L_{\mu} = {\sigma_{\mu}}^{-}$.

If the average excitation number of the system is unchanged, it is a simple solution phase process, and only the quantum coherence decays exponentially with time, then $L_{\mu} = {\sigma_{\mu}}^{+}{\sigma_{\mu}}^{-}$.

If the coupling heat storage temperature with the system tends to be infinite and the coupling strength tends to zero resulting noise, then $L_{\mu} = {\sigma_{\mu}}^{-}$ and $L_{\mu} = {\sigma_{\mu}}^{+}$.

If in Pauli and de-sharpening noise environment, then $L_{\mu} = \sigma_{\mu}(i = x, y, z)$. Different quantum noise channel models (Pauli channel, bit-flipping channel, amplitude damping channel, depolarization channel, etc.) are used to simulate the noise environment of a physical system. Among them, the bit-flipping channel will with probability $1 - P$ flip from $|0\rangle$ to $|1\rangle$ (or vice versa). We express the generation of noise as $M_0 = \sqrt{P}\begin{pmatrix} 1 & 0 \\ 0 & 1 \end{pmatrix}, M_1 = \sqrt{1-P}\sigma_x = \sqrt{1-P}\begin{pmatrix} 0 & 1 \\ 1 & 0 \end{pmatrix}$. In the depolarization channel, the quantum bits are depolarized with probability P, and the entanglement evolution of this process is obvious, while the probability of $1 - P$ remains unchanged, and then the quantum system is $\rho_{\text{sys}}(t) = \frac{PI}{2} + (1-P)\rho$. For d-dimensional quantum system, the depolarization channel is still replaced by a completely mixed state with probability P, and the model is $\rho_{\text{sys}}(t) = \frac{PI}{2} + (1-P)\rho$. The depolarized noise density matrix is completely unpolarized, and the *kraus* model is $E_1 = \sqrt{1-3P/4}, E_2 = \sqrt{P/4}\sigma_x, E_3 = \sqrt{P/4}\sigma_y,\ E_4 = \sqrt{P/4}\sigma_z$. Amplitude damping channels can change the phase with probability P through *kraus* operation: $|1\rangle \to -|1\rangle, E_1 = \sqrt{1-P}I, E_\ni = \sqrt{P}\sigma_z$. The following focus on describing the entanglement evolution process based on Pau*li* noise in local common mode.

3.1.3.1 Entanglement Evolution Under Pau*li* σ_x

When the mixed entangled state is under Pau*li* σ_x noise, we get the Lindblad operator $(L_{1,x} = \sqrt{k}\sigma_x \otimes I,\quad L_{2,x} = I \otimes \sqrt{k}\sigma_x)$. In the main equation $\frac{d\rho(t)}{dt} = \sum_\mu \left(L_\mu\rho(t)L_\mu^\dagger - \frac{1}{2}L_\mu^\dagger L_\mu\rho(t) - \frac{1}{2}\rho(t)L_\mu^\dagger L_\mu\right)$, there are two types of density matrix when time is t. The first type is described as follows:

$$\rho_{Mi}(0) \to \rho_x^{in,Mi}(t) = \begin{bmatrix} \rho_{11}^{in,M1(2)} & 0 & 0 & \rho_{14}^{in,M1(2)} \\ 0 & \rho_{22}^{in,M1(2)} & \rho_{23}^{in,M1(2)} & 0 \\ 0 & \rho_{32}^{in,M1(2)} & \rho_{33}^{in,M1(2)} & 0 \\ \rho_{41}^{in,M1(2)} & 0 & 0 & \rho_{44}^{in,M1(2)} \end{bmatrix}. \tag{3.34}$$

In Eq. (3.34),

$$\rho_{11}^{in,M_1} = \frac{\left[1-e^{-4kt}+2\alpha_{M1}\left(e^{-4kt}+e^{-2kt}\right)\right]}{4},\ \rho_{11}^{in,M_2} = \frac{\left[1-e^{-4kt}+2\alpha_{M2}\left(e^{-4kt}+e^{-2kt}\right)\right]}{4},$$

$$\rho_{22}^{in,M1(2)} = \rho_{33}^{in,M1(2)} = \frac{e^{-4kt}\left(e^{4kt}+1-2\alpha_{M1(2)}\right)}{4},\ \rho_{44}^{in,M2} = \frac{\left[1-e^{-4kt}+2\alpha_{M2}\left(e^{-4kt}+e^{-2kt}\right)\right]}{4},$$

$$\rho_{14}^{in,M1(2)} = \rho_{41}^{in,M1(2)=\left[e^{-4kt}\left(1-\alpha_{M1(2)}\right)\left(e^{4kt}-1\right)\right]/4},\ \rho_{44}^{in,M1} = \frac{\left[1-e^{-4kt}+2\alpha_{M1}\left(e^{-4kt}-e^{-2kt}\right)\right]}{4},$$

and $$\rho_{23}^{in,M1(2)} = \rho_{32}^{in,M1(2)} = \frac{e^{-4kt}\left(1-\alpha_{M1(2)}\right)\left(e^{4kt}+1\right)}{4}$$

represent the density matrix element under Pau*li* σ_x.

Thus, we get the concurrency when time is t, which is described as follows:

$$C_x^{in,Mi} = m\left\{0, 2\sqrt{\rho_{23}^{Mi}\rho_{32}^{Mi}} - 2\sqrt{\rho_{11}^{Mi}\rho_{44}^{Mi}}\right\} = \frac{1}{2}e^{-4kt}q \tag{3.35}$$

In Eq. (3.35),

$$q = \left|(\alpha_{Mi} - 1)(e^{4kt} + 1)\right| - e^{4kt}\sqrt{e^{-8kt}(e^{4kt} - 1)(e^{4kt} + 4\alpha_{Mi} - 1 - 4\alpha_{Mi}^2)}.$$

The second type is described below:

$$\rho_{Ni}(0) \rightarrow \rho_x^{in,Ni}(t) = \begin{bmatrix} \rho_{11}^{in,N1(2)} & 0 & 0 & \rho_{14}^{in,N1(2)} \\ 0 & \rho_{22}^{in,N1(2)} & \rho_{23}^{in,N1(2)} & 0 \\ 0 & \rho_{32}^{in,N1(2)} & \rho_{33}^{in,N1(2)} & 0 \\ \rho_{41}^{in,N1(2)} & 0 & 0 & \rho_{44}^{in,N1(2)} \end{bmatrix} \tag{3.36}$$

In Formula (3.36),

$$\rho_{11}^{in,N1} = = \frac{e^{-4kt}(e^{4kt} + 1 + e^{2kt}\alpha_{N1})}{4},\ \rho_{11}^{in,N2} = \frac{e^{-4kt}(e^{4kt} + 1 + e^{2kt}\alpha_{N1})}{4},$$

$$\rho_{22}^{in,N1(2)} = \rho_{33}^{in,N1(2)} = \frac{e^{2k,N1}(2)}{4},\ \rho_{14}^{in,N1(2)} = \rho_{41}^{inN1(2)} = \frac{e^{-4kt}(2 - a_{N1(2)}(e^{4kt} + 1)}{4},$$

$$\rho_{23}^{in,N1(2)} = \rho_{32}^{in,N1(2)} = \frac{e^{-4kt}(2 - \alpha_{N1(2)})(e^{4kt} - 1)}{4},\ \rho_{44}^{in,N1} = \frac{e^{-4kt}(e^{4kt} + 1 - e^{2kt}\alpha_{N1})}{4},$$

and $\rho_{44}^{in,N2} = \frac{e^{-4kt}(e^{4kt} + 1 + e^{2kt}\alpha_{N2})}{4}$

represent the density matrix element under Pau*li* σ_x

The concurrency when time is t can be obtained through the calculation

$$\begin{aligned} C_{xi}^{in,Ni} &= m\left\{0, 2\sqrt{\rho_{14}^{Ni}\rho_{41}^{Ni}}\right\} - 2\sqrt{\rho_{22}^{Ni}\rho_{33}^{Ni}}\Bigg\} \\ &= \frac{1}{4}\left\{e^{-4kt}\left|(1 + e^{4kt})(\alpha_{N_i} - 2)\right| - 2\sqrt{e^{-8kt}(e^{4kt} - 1)^2}\right\} \end{aligned} \tag{3.37}$$

It can be seen that the concurrency of the system under locally independent Pauli noise is a function related to stacking angle γ, purity of initial conditions r, and time kt.

Under local independent Pauli noise and when $r = 1$, the system is in Bell pure state. The evolution of system concurrency is shown in Fig. 3.17.

As can be seen from the analysis in Fig. 3.17, the degree of concurrency changes periodically with the stacking angle γ and always decays in finite time and eventually goes to zero. When $\gamma = (1 + 2n)\frac{\pi}{4}, n = 0, 1, 3, \ldots\ldots$, the system is in the maximum entangled Bell state, the concurrency will gradually decline exponentially, and when

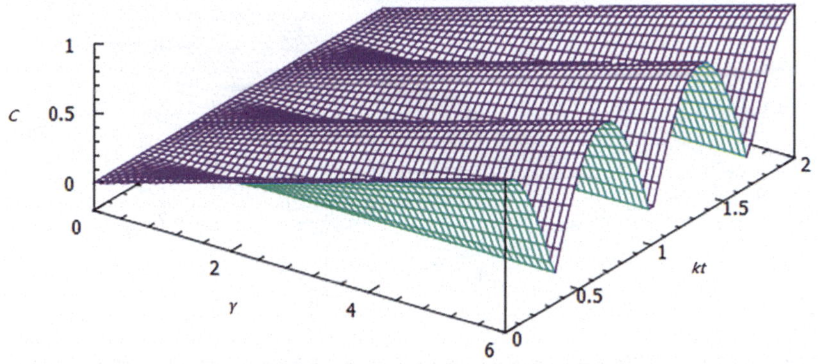

Fig. 3.17 System concurrency evolution under local independent Pauli noise when $r = 1$

time is ∞, the concurrency decays to zero. When $r < 1$,the system was in a mixed state. By the formula

$$C\left(\rho^{x,in}(t)\right) = 2m\left\{0, \sqrt{\rho_{14}^{x,in}\rho_{41}^{x,in}} - \sqrt{\rho_{22}^{x,in}\rho_{33}^{x,in}}\right\} = \frac{r\left(1 + e^{-4kt}\right)|\sin 2\gamma|}{2} - \frac{\left|re^{-4kt} - 1\right|}{2},$$

we can know the time kt when the concurrency decays to zero and the concurrency γ is related to r.

In locally independent Pauli noise, the system concurrency evolves as shown in Fig. 3.18.

As can be seen from Fig. 3.18, the degree of concurrency also changes periodically with γ and exists in a certain time interval.

3.1.3.2 Entanglement Evolution Under Pau*li* σ_y

When the mixed entangled state is under Pau*li* σ_y noise, we get the Lindblad operator $\left(L_{1,y} = \sqrt{k}\sigma_y \otimes I,\ L_{2,y} = I \otimes \sqrt{k}\sigma_y\right)$. In the main equation, there are two types of density matrix when time is t. The first type is described as follows:

$$\rho_{Mi}(0) \rightarrow \xi_y^{in,Mi}(t) = \begin{bmatrix} \xi_{11}^{in,M1(2)} & 0 & 0 & \xi_{14}^{in,M1(2)} \\ 0 & \xi_{22}^{in,M1(2)} & \xi_{23}^{in,M1(2)} & 0 \\ 0 & \xi_{32}^{in,M1(2)} & \xi_{33}^{in,M1(2)} & 0 \\ \xi_{41}^{in,M1(2)} & 0 & 0 & \xi_{44}^{in,M1(2)} \end{bmatrix}. \tag{3.38}$$

In Eq. (3.38),

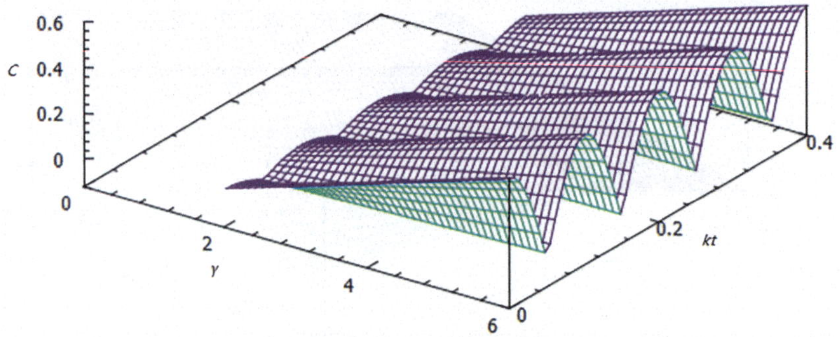

Fig. 3.18 Evolution of system concurrency under locally independent Pauli noise

$\xi_{44}^{in,M1(2)}=\rho_{44}^{in,M1(2)}$, $\xi_{11}^{in,M1(2)}=\rho_{11}^{in,M1(2)}$, $\xi_{22}^{in,M1(2)}=\rho_{22}^{in,M1(2)}$, $\xi_{33}^{in,M1(2)}=\rho_{33}^{in,M1(2)}$, $\xi_{23}^{in,M1(2)}=\rho_{23}^{in,M1(2)}$, $\xi_{32}^{in,M1(2)}=\rho_{32}^{in,M1(2)}$, and $\xi_{14}^{in,M1(2)}=\xi_{41}^{in,M1(2)}=-\rho_{14}^{in,M1(2)}=-\rho_{41}^{in,M1(2)}$. $\xi_{ij}^{in,M1(2)}(i(j)=1,2,3,4)$ represents the density matrix element in row i and column j taking $\rho_{M1(2)}(0)$ as the initial state under σ_y noise, and the concurrency can be calculated when time is t:

$$C_y^{in,Mi}=C_x^{in,Mi}. \tag{3.39}$$

The second type is described below:

$$\rho_{Ni}(0)\rightarrow\xi_y^{in,Ni}(t)=\begin{bmatrix}\xi_{11}^{in,N1(2)} & 0 & 0 & \xi_{14}^{in,N1(2)}\\ 0 & \xi_{22}^{in,N1(2)} & \xi_{23}^{in,N1(2)} & 0\\ 0 & \xi_{32}^{in,N1(2)} & \xi_{33}^{in,N1(2)} & 0\\ \xi_{41}^{in,N1(2)} & 0 & 0 & \xi_{44}^{in,N1(2)}\end{bmatrix}. \tag{3.40}$$

In Eq. (3.40), $\xi_{23}^{in,N1(2)}=\xi_{32}^{in,N1(2)}=-\rho_{23}^{in,N1(2)}=-\rho_{32}^{in,N1(2)}$, $\xi_{11}^{in,N1(2)}=\rho_{11}^{in,N1(2)}$, $\xi_{14}^{in,N1(2)}=\rho_{41}^{in,N1(2)}$, $\xi_{22}^{in,N1(2)}=\rho_{22}^{in,N1(2)}$, $\xi_{33}^{in,N1(2)}=\rho_{33}^{in,N1(2)}$, $\xi_{41}^{in,N1(2)}=\rho_{41}^{in,N1(2)}$, and $\xi_{44}^{in,N1(2)}=\rho_{44}^{in,N1(2)}$. $\xi_{ij}^{in,N1(2)}(i(j)=1,2,3,4)$ represents the density matrix element in row i and column j taking $\rho_{N1(2)}(0)$ as the initial state under σ_y noise, and the concurrency can be calculated when time is t:

$$C_y^{in,Ni}=C_x^{in,Ni}. \tag{3.41}$$

As shown in Fig. 3.19 regarding the concurrency in a local common noise environment, the concurrency is determined by α_{Ni}. If α_{Ni} is known, the initial value of concurrency will be unchanged, but with the increase of purity, the entanglement degree of the system will be less, and the entanglement of $\rho Ni(0)$ in the common σ_y noise will be reduced with the increase of pure state. It can be seen

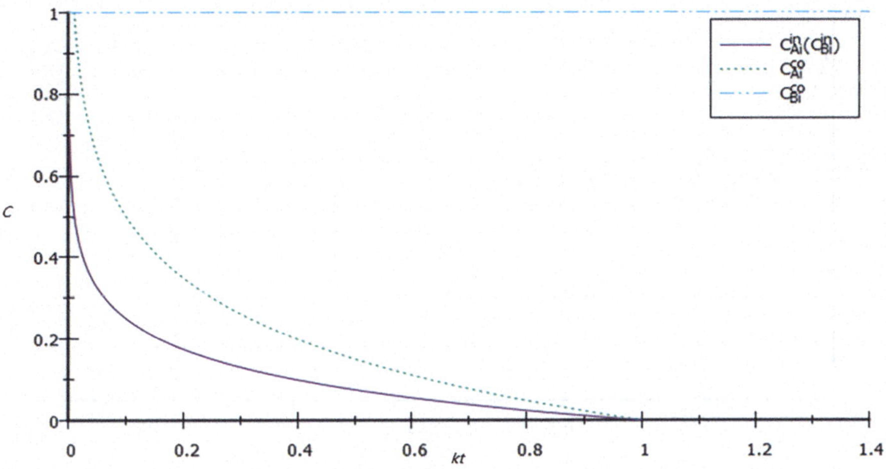

Fig. 3.19 The concurrency when $\alpha_{Mi} = 0$, $\alpha_{Ni} = 0$

that in this way the problem of entanglement death is effectively avoided and the maximum coherence is maintained.

Next, we analyze and discuss the evolution of concurrency with variables *kt* under different value of α_M and α_N, as shown in Fig. 3.19.

As shown in Fig. 3.19, there was no sudden death in the concurrency of the system when $\alpha_{Mi} = 0$ and $\alpha_{Ni} = 0$. In the maximum entangled state which is ($|10\rangle+|01\rangle$)($\langle 10|+\langle 01|$)/2 in common noise, the degree of concurrency decreases gradually. The concurrency of the maximum entangled state ($|11\rangle+|00\rangle$)($\langle 11|+\langle 00|$)/2 is always 1, which shows that the maximum entangled state can resist the influence of common noise and maintain the coherence of quantum.

When $\alpha_{Mi} = 0.5$, $\alpha_{Ni} = 1$, the system is in the mixed state and the initial concurrency decreases as shown in Fig. 3.20.

As shown in Fig. 3.20, if $\alpha_{Mi} = 0.5$, $\alpha_{Ni} = 1$, in other words, the concurrency in the initial state is 0.5, and the entanglement value in the common noise will always be the same, and in other cases entanglement and death will occur over time. When the initial state is ($|10\rangle+|01\rangle$)($\langle 10|+\langle 01|$)/2, the attenuation rate in local independent noise is slower than that in common noise.

3.1.3.3 Entanglement Evolution Under Pau*li* σ_z

When the mixed entangled state is under Pau*li* σ_z noise, we get the Lindblad operator $\left(L_{1,z} = \sqrt{k}\sigma_z \otimes I,\ L_{2,z} = I \otimes \sqrt{k}\sigma_z\right)$. In the main equation, there are two types of density matrix when time is *t*. The first type is described as follows:

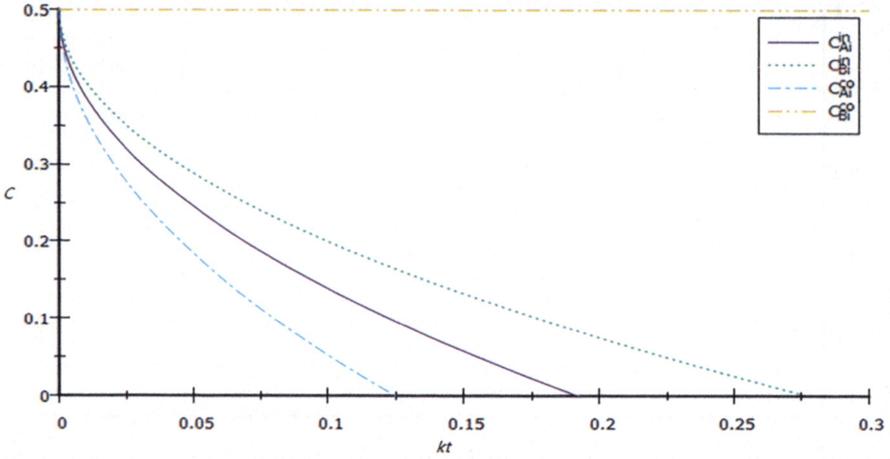

Fig. 3.20 The concurrency when $\alpha_{Mi} = 0.5$, $\alpha_{Ni} = 1$

$$\rho_{Mi}(0) \rightarrow \mathrm{F}_z^{in,Mi}(t) = \begin{bmatrix} \mathrm{F}_{11}^{in,M1(2)} & 0 & 0 & \mathrm{F}_{14}^{in,M1(2)} \\ 0 & \mathrm{F}_{22}^{in,M1(2)} & \mathrm{F}_{23}^{in,M1(2)} & 0 \\ 0 & \mathrm{F}_{32}^{in,M1(2)} & \mathrm{F}_{33}^{in,M1(2)} & 0 \\ \mathrm{F}_{41}^{in,M1(2)} & 0 & 0 & \mathrm{F}_{44}^{in,M1(2)} \end{bmatrix}. \tag{3.42}$$

In Eq. (3.42), $\mathrm{F}_{11}^{in,M1} = \alpha_{A1}$, $\mathrm{F}_{11}^{in,M2} = \mathrm{F}_{44}^{in,M1} = 0$, $\mathrm{F}_{44}^{in,M2} = \alpha_{M2}$, $\mathrm{F}_{22}^{in,M1(2)} = \mathrm{F}_{33}^{in,M1(2)} = \frac{1-\alpha^{A1(2)}}{2}$, and $\mathrm{F}_{23}^{in,M1(2)} = \mathrm{F}_{32}^{in,M1(2)} = \frac{e^{-4kt}\left(1-\alpha^{M1(2)}\right)}{2}$. $\mathrm{F}_{ij}^{in,M1(2)}(i(j) = 1, 2, 3, 4)$ represents the density matrix element in row i and column j taking $\rho_{M1(2)}(0)$ as the initial state under σ_z noise, and the concurrency can be calculated when time is t:

$$\begin{aligned} C_z^{in,Mi} &= m\left\{0, 2\sqrt{\sum_{23}^{Mi}\sum_{32}^{Mi} - 2\sum_{11}^{Mi}\sum_{44}^{Mi}}\right\} \\ &= (1-\alpha_{Mi})e^{-4kt}. \end{aligned} \tag{3.43}$$

The second type is described below:

$$\rho_{Ni}(0) \rightarrow \mathrm{F}_z^{in,Ni}(t) = \begin{bmatrix} \mathrm{F}_{11}^{in,N1(2)} & 0 & 0 & \mathrm{F}_{14}^{in,N1(2)} \\ 0 & 0 & 0 & 0 \\ 0 & 0 & 0 & 0 \\ \mathrm{F}_{41}^{in,N1(2)} & 0 & 0 & \mathrm{F}_{44}^{in,N1(2)} \end{bmatrix}. \tag{3.44}$$

In Eq. (3.44), $\mathrm{F}_{11}^{in,N2} = \frac{2+\alpha_{N1}}{4}$, $\mathrm{F}_{44}^{in,N2} = \frac{2-\alpha_{N1}}{4}$, $\mathrm{F}_{14}^{in,N1(2)} = \mathrm{F}_{41}^{in,N1(2)} = \frac{e^{-4kt}\left(2-\alpha^{N1(2)}\right)}{4}$, $\mathrm{F}_{11}^{in,N2} = \frac{2-\alpha_{N2}}{4}$, and $\mathrm{F}_{44}^{in,N2} = \frac{2+\alpha_{N2}}{4}$. $\mathrm{F}_{ii}^{in,N1(2)}(i(j) = 1, 2, 3, 4)$ represents the density matrix element in

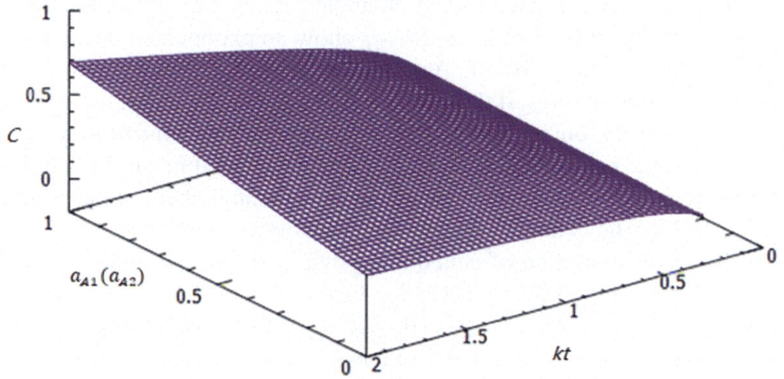

(a) Concurrency of $FMi(t)$

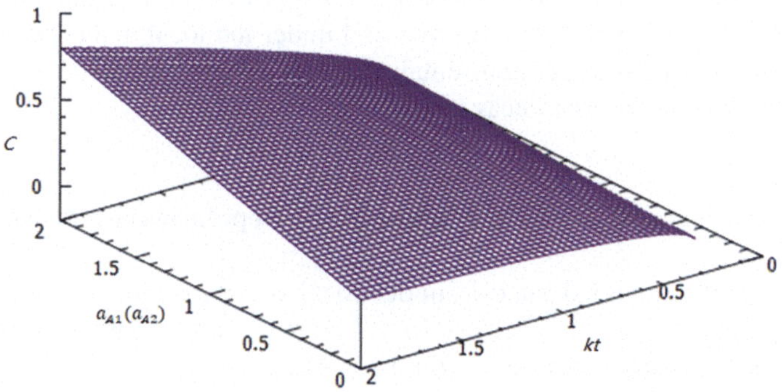

(b) Concurrency of $F_{Ni}N(t)$

Fig. 3.21 Concurrency of entanglement evolution. (**a**) Concurrency of $F_{Mi(t)}$. (**b**) Concurrency of $F_{Ni}N(t)$

row i and column j taking $\rho_{N1(2)}(0)$ as the initial state under σ_z noise, and the concurrency can be calculated when time is t:

$$\begin{aligned} C_z^{in,Ni} &= m\left\{0, 2\sqrt{\sum_{14}^{Ni}\sum_{41}^{Ni}} - 2\sqrt{\sum_{22}^{Ni}\sum_{33}^{Ni}}\right\}. \\ &= (2-\alpha_{Ni})e^{-4kt} \end{aligned} \tag{3.45}$$

Entanglement evolution of mixed entangled states under local common noise is shown in Fig. 3.21:

As shown in Fig. 3.21, when mixed entangled states are subject to common σ_z noise, both entanglements of $F_{Mi(t)}$ and $F_{Ni(t)}$ show an exponential decay trend, and entanglement death will not occur. As shown in Fig. 3.21a, the concurrency of a system is related to purity α_{Mi}. If the purity α_{Mi} decreases gradually, the concurrency will decrease gradually, but when α_{Mi} is in the given case, the concurrency of $F_{Mi(t)}$ is invariable. Thus, the entanglement death problem is avoided in a noise environment by the entanglement of $\rho_{Mi(0)}$ state which keeps the entanglement of the system very well. In Fig. 3.21b, the concurrency of $F_{Ni(t)}$ state decays gradually.

We analyzed the evolution of concurrency with variables kt under different α_{Mi} and α_N, as shown in Fig. 3.22.

As shown in Fig. 3.22a, if $\alpha_{Mi} = 0$, $\alpha_{Ni} = 0$, the concurrency of maximum entangled state $(|10\rangle + |01\rangle)(\langle 10| + \langle 01|)/2$ is 1 which is at maximum. That means it can resist common noise σ_z very well. And the concurrency of $(|00\rangle + |11\rangle)(\langle 00| + \langle 11|)/2$ both appears as exponential form and gradually decays in the local independent or common noise.

As shown in Fig. 3.22b, if $\alpha_{Mi} = 0.5$, $\alpha_{Ni} = 1$, the concurrency of the maximum entangled state $(|10\rangle + |01\rangle)(\langle 10| + \langle 01|)/2$ is kept at 0.5 under the local common noise and exhibits an exponential decay trend under the local independent noise σ_z. The concurrency of the maximum entangled state $(|00\rangle + |11\rangle)(\langle 00| + \langle 11|)/2$ shows a trend of exponential decay in both kinds of noise.

3.1.3.4 Entanglement Evolution Under Local Depolarization Noise

When the mixed entangled state is under local depolarization noise, we get the Lindblad operator ($L_{1,x} + L_{2,x}$, $L_{1,y} + L_{2,y}$, $L_{1,z} + L_{2,z}$). In the main equation, there are two types of density matrix when time is t. The first type is described as follows:

$$\rho_{Mi}(0) \rightarrow \Xi_d^{co,Mi}(t) = \begin{bmatrix} \Xi_{11}^{co,M1(2)} & 0 & 0 & \Xi_{14}^{co,M1(2)} \\ 0 & \Xi_{22}^{co,M1(2)} & \Xi_{23}^{co,M1(2)} & 0 \\ 0 & \Xi_{32}^{co,M1(2)} & \Xi_{33}^{co,M1(2)} & 0 \\ \Xi_{41}^{co,M1(2)} & 0 & 0 & \Xi_{44}^{co,M1(2)} \end{bmatrix}. \tag{3.46}$$

In Eq. (3.46), $\Xi_{11}^{co,M1} = \left[2\left(1-e^{-12kt}\right) + 3\alpha_{M1}\left(e^{-12kt} + e^{-4kt}\right)\right]/6$, $\Xi_{14}^{co,M1(2)} = \Xi_{41}^{co,M1(2)} = 0$, $\Xi_{11}^{co,M2} = \left[2\left(1-e^{-12kt}\right) + 3\alpha_{M2}\left(e^{-12kt} + e^{-4kt}\right)\right]/6$, $\Xi_{44}^{co,M1} = \left[2\left(1-e^{-12kt}\right) + 3\alpha_{M1}\left(e^{-12kt} + e^{-4kt}\right)\right]/6$, $\Xi_{22}^{co,M1(2)} = \Xi_{33}^{co,M1(2)} = \Xi_{23}^{co,M1(2)} = \Xi_{32}^{co,M1(2)} = e^{-12kt}\left(e^{12kt} + 2 - 3\alpha_{M1(2)}\right)/6$, and $\Xi_{44}^{co,M2} = \left[2\left(1-e^{-12kt}\right) + 3\alpha_{M2}\left(e^{-12kt} + e^{-4kt}\right)\right]/6$. $\Xi_{ij}^{co,M1(2)}(i(j) = 1,2,3,4)$ represents the density matrix element in row i and column j taking $\rho_{M1(2)}(0)$ as the initial state under σ_z noise, and the concurrency can be calculated when time is t:

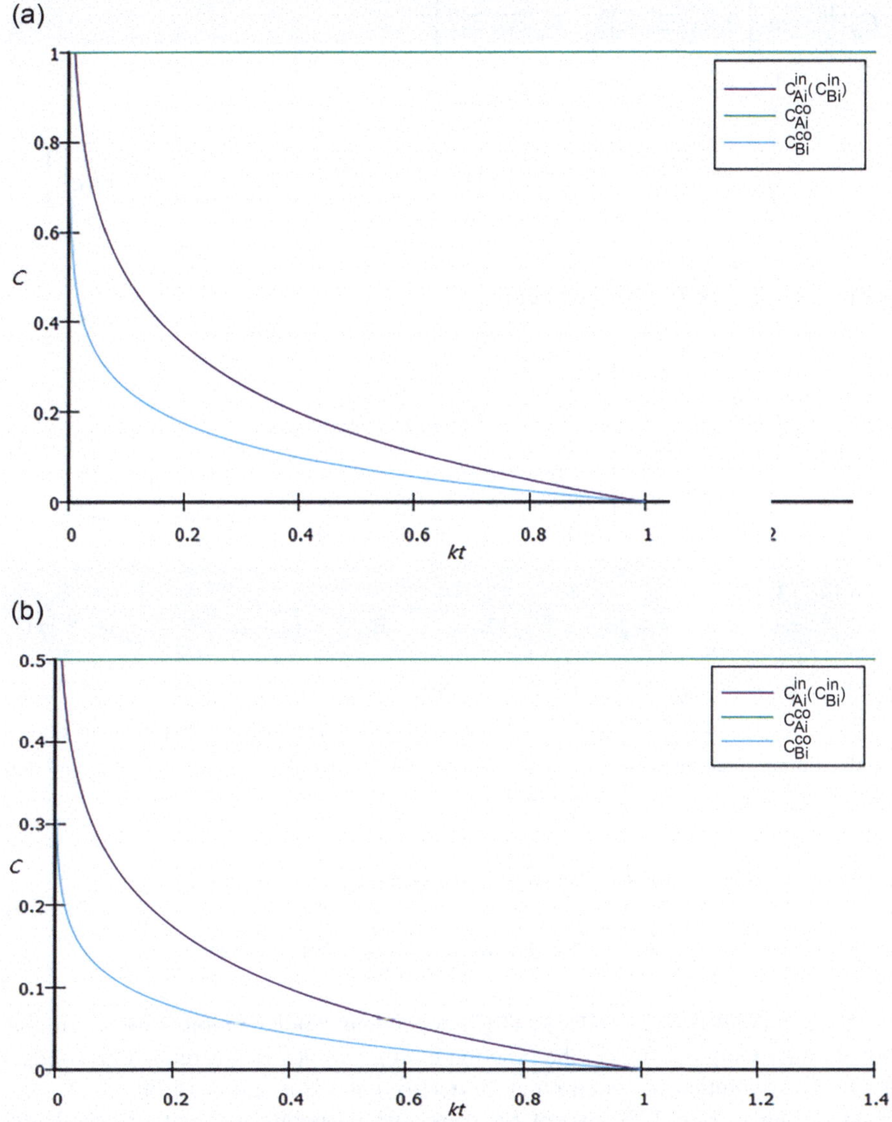

Fig. 3.22 (**a**) Concurrency when $\alpha_{Mi} = 0$, $\alpha_{Ni} = 0$. (**b**) Concurrency when $\alpha_{Mi} = 0.5$, $\alpha_{Ni} = 1$

$$C_d^{co,Mi} = m\left\{0, 2\sqrt{\Xi_{23}^{Mi}\Xi_{32}^{Mi}} - 2\sqrt{\Xi_{11}^{Mi}\Xi_{44}^{Mi}}\right\}$$
$$= \frac{1}{3}e^{-12kt}\left\{\left|2+e^{12kt}-3\alpha_{Mi}\right| - e^{-12kt}\right.$$
$$\left.\sqrt{e^{-24kt}\left[4(-1+e^{12kt})^2+12(-1+e^{12kt})\alpha_{Mi}-9(-1+e^{16kt})\alpha_{Mi}^2\right]}\right\}. \tag{3.47}$$

The second type is described below:

$$\rho_{Ni}(0) \to \Xi_d^{co,Ni}(t) = \begin{bmatrix} \Xi_{11}^{co,N1(2)} & 0 & 0 & \Xi_{14}^{co,N1(2)} \\ 0 & \Xi_{22,N1(2)}^{co} & \Xi_{23,N1(2)}^{co,} & 0 \\ 0 & \Xi_{32}^{c,N1(2)} & \Xi_{33}^{co,N1(2)} & 0 \\ \Xi_{41}^{co,N1(2)} & 0 & 0 & \Xi_{44}^{co,N1(2)} \end{bmatrix}. \tag{3.48}$$

In Eq. (3.48), $\Xi_{11}^{co,N1} = (4+2e^{-12kt}+3\alpha_{N1}e^{-4kt})/12$, $\Xi_{11}^{co,N2} = (4+2e^{-12kt}-3\alpha_{N2}e^{-4kt})/12$, $\Xi_{44}^{co,N1} = (4+2e^{-12kt}-3\alpha_{N1}e^{-4kt})/12$, $\Xi_{44}^{co,N2} = 4+2e^{-12kt}+3\alpha_{N2}e^{-4kt}/12$, $\Xi_{22}^{co,N1(2)} = \Xi_{33}^{co,N1(2)} = \Xi_{23}^{co,N1(2)} = \Xi_{32}^{co,N1(2)} = e^{-12kt}(e^{12kt}-1)/6$, and $\Xi_{14}^{co,N1(2)} = \Xi_{41}^{co,N1(2)} = e^{-12kt}(2-\alpha_{N1(2)})/4$. $\Xi_{ij}^{co,N1(2)}(i(j)=1,2,3,4)$ represents the density matrix element in row i and column j taking $\rho_{N1(2)}(0)$ as the initial state under depolarization noise, and the concurrency can be calculated when time is t:

$$C_d^{co,Ni} = m\left\{0, 2\sqrt{\Xi_{14}^{Ni}\Xi_{41}^{Ni}} - 2\sqrt{\Xi_{22}^{Ni}\Xi_{33}^{Ni}}\right\}$$
$$= \frac{1}{2}e^{-12kt}\left|-2+\alpha_{Ni}\right| - \frac{1}{3}\sqrt{e^{-24kt}(-1+e^{12kt})^2}. \tag{3.49}$$

Mixed entangled states under locally independent depolarization noise are similar to that under Pauli σ_x. We get the function of concurrency that is related to $\alpha_{Mi}(\alpha_{Ni})$ and kt. The evolution of concurrency of $\alpha_{Mi}(\alpha_{Ni})$ and kt is shown in Fig. 3.23.

As shown in Fig. 3.23, except for some special points $\alpha_{Mi} = 1$, $\alpha_{Ni} = 2$, the entanglement will disappear in finite time in the system taking $\rho_{Mi}(0)(i = 1,2)$ or $\rho_{Ni}(0)(i = 1,2)$ as the initial state. When $\alpha_{Mi} = 1$, the system is in a separate state $|00\rangle$. When $\alpha_{Ni} = 2$, the system is in a separate state $|11\rangle$ and the concurrency state is 0. When $\alpha_{Mi} = 0$, $\alpha_{Ni} = 0$, the system is in maximum entangled state $(|10\rangle + |01\rangle)(\langle 10| + \langle 01|)/2$ or $(|00\rangle + |11\rangle)(\langle 00| + \langle 11|)/2$, and at the same time, the concurrency of the system disappeared in a finite time because the Lindblad operator is not commutative, that is to say, $[L_{1,x} + L_{2,x}] \neq 0$. The evolution of the concurrency in different parameters α_{Mi}, α_{Ni} with kt is shown in Fig. 3.24.

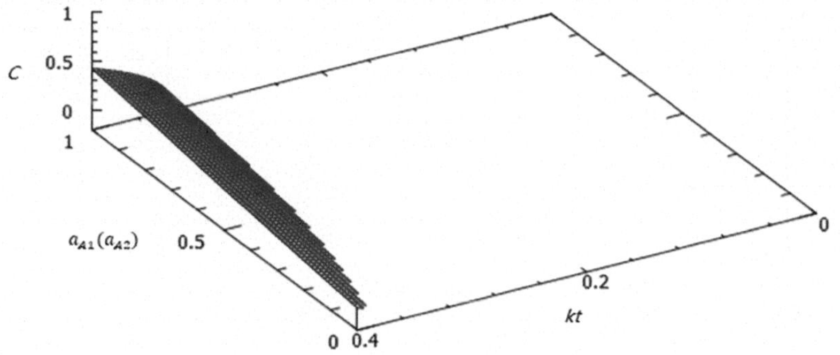

(a)

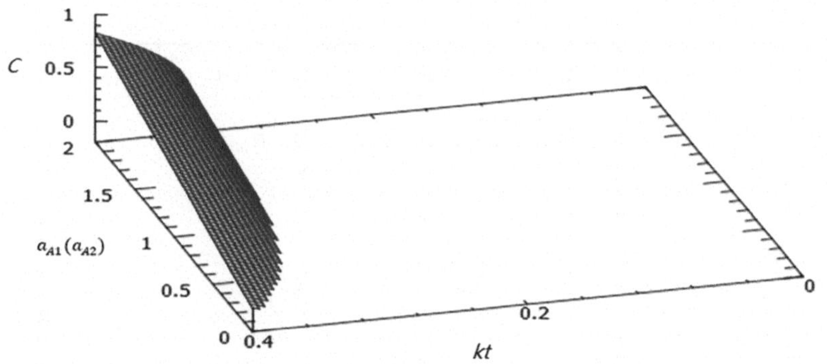

(b)

Fig. 3.23 Diagram showing the evolution of concurrency of $\Xi_{Mi(t)}$ $\Xi_{Ni(t)}$ and kt

As shown in Fig. 3.24a, the concurrency attenuates to zero when $\alpha_{M_i} = 0, \alpha_{V_i} = 0$ and attenuates faster under the local common noise than the local independent noise.

As shown in Fig. 3.24b, when $\alpha_{Mi} = 0.5$, $\alpha_{Ni} = 1$, the concurrency at the initial moment is 0.5. The common phenomenon is that the concurrency in the local common environment attenuates faster than that in the local independent environment. In comparison, the attenuation rate of the initial state $\rho_{Mi}(0)$ in the local independent environment is the fastest, while the attenuation rate of the initial state $\rho_{Ni}(0)$ in the local independent environment is the slowest which describes the entanglement evolution process under the depolarization noise well.

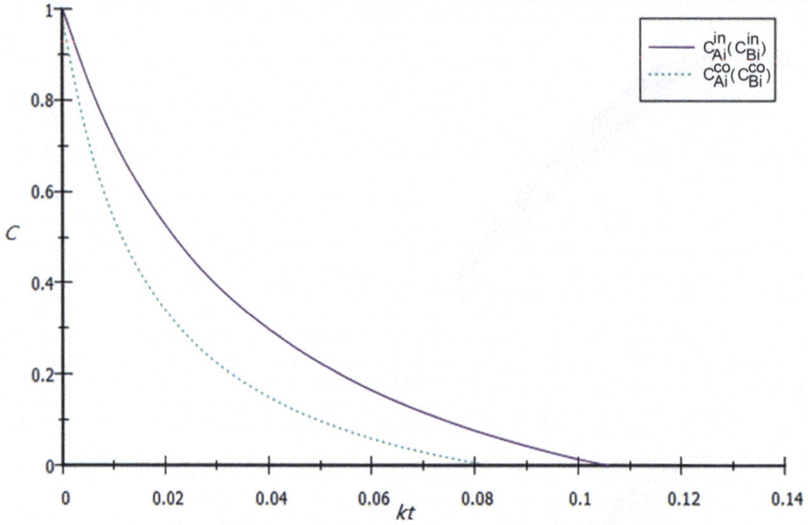

(a) Concurrency when $\alpha_{Mi} = 0,\ \alpha_{Ni} = 0$

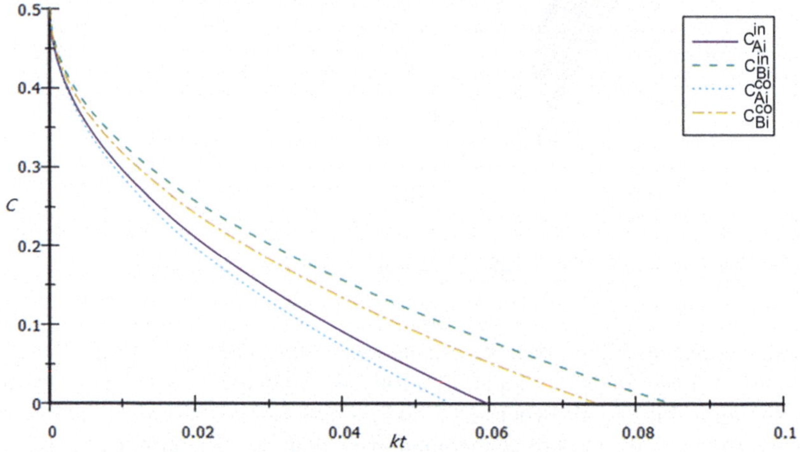

(b) Concurrency when $\alpha_{Mi} = 0.5, \alpha_{Ni} = 1$

Fig. 3.24 Evolution of concurrency with variables kt under different parameters α_{Mi} and α_{Ni}. (**a**) Concurrency when $\alpha_{Mi} = 0$, $\alpha_{Ni} = 0$. (**b**) Concurrency when $\alpha_{Mi} = 0.5$, $\alpha_{Ni} = 1$

3.2 Building an Immune Noise Model

Based on characterization of quantum entanglement evolution and unitary transformation in Sect. 3.1, we analyzed the noise characteristics of the real physical system coupling with the environment, build the dynamic switching model immunizing different noise, and also build immune noise model based on DFS and density matrix which is robust. They are the core of a unified framework of high-fidelity entanglement quantum teleportation channel.

3.2.1 *Immune Noise Model Based on Density Matrix*

3.2.1.1 Properties

In order to simultaneously achieve high fidelity and security of the channel, the properties of the immune noise model designed by us are described as follows:

The interaction of quantum qubits under noise is represented by the noise operator interacting with each other in Hilbert space. E_k represents one type of noise in *Kraus* operation which is used to keep track of this operation (probability conservation) and satisfy the condition $\sum_{i=1}^{n} E_j E_j = I$. I is the identity matrix in the Hilbert space. The total density matrix $\rho = \rho_{in} \otimes \rho_{ch}$ describes the initial state, which can change each qubit independently according to the noise type. By using the noise operator on the qubit k and the density matrix $\rho_k \rightarrow \rho'_k = E_{j=1}^{n} E_j \rho_k E_j^{\dagger}$, the total density matrix after the action of multiple noise sources is

$$\rho' = \sum_{j=1}^{n_1} E_i(P_I) \left[\sum_{j=1}^{n'} F_j(P_A) \left(\sum_{K=1}^{n_B} G_k(P_B) \rho G_k^{\dagger}(P_B) \right) E_j^{\dagger}(P_A) \right] E_i^{\dagger}(P_I). \quad (3.50)$$

For the probability of each quantum channel state occurring in quantum noise channel, we define the fidelity as $F_j = \left\langle \psi \middle| \rho_{B_j} \middle| \psi \right\rangle$ and average fidelity as $\overline{F} = \sum^4 T_r \left[P_j^{\varphi} \rho \right] T_r \left[\rho_{in} \rho'_{B_j} \right]$ to measure the effectiveness and safety of immune models. In order to quantify and measure the impact of noise, the optimal efficiency $\overline{\langle F_{X,\phi,Y} \rangle}$ is defined where X is the noise channels affecting qubits, ϕ is the symbol indicating that Alice in the quantum channel does not suffer from noise, and Y denotes the qubit of Bob subjected to noise.

Specifically, we can analyze it in three cases:

1. When quantum noise channels (bit-flipping channel, depolarization channel, amplitude damping channel, Pauli channel, etc.) are free from noise ($P_A = P_B = 0$) and only Alice's input quantum is in noise environment ($P_I \neq 0$), the average noise fidelity and optimal efficiency value of each quantum channel type are described and quantified. Furthermore, measuring the condition that the input one Alice's qubit is subjected to one of the noises and Bob's qubit may be in some other noises and describing and quantifying the average fidelity and optimal efficiency under different conditions. When the noise cannot be avoided, the average fidelity and optimal efficiency of the qubit in the teleportation process under different noises are further analyzed.

2. When all of Alice's qubits are exposed to the same noise type ($P_A = P_B = P$), the average fidelity and optimal efficiency are analyzed and quantified under scenarios where Alice's qubit experiences one noise type while Bob's qubit undergoes multiple noise types.
3. When the channel qubits are exposed to identical noise types—i.e., the qubit experiences noise during transmission from Alice to Bob—the input qubit may undergo one or more distinct noise types. The average fidelity and optimal efficiency are respectively analyzed and quantified under these varying conditions.

3.2.1.2 Noise Modeling

The correlation of a single particle in a noisy environment is represented by quantum operations in Hilbert space. Operators E_j represent certain types of noise which are often called *Kraus* operators and used to track and hold operations (probability conservation) [7]. They satisfy the following conditions:

$$\sum_{i=1}^{n} E_j^{\dagger} E_j = I \tag{3.51}$$

where $1 < n < 4$ and I is the identity matrix acting on Hilbert space. k represents the effect of noise on qubits, and we get the density matrix ρ_k:

$$\rho_k \rightarrow \rho_k = \sum_{j=1}^{n} E_j \rho_k E_j^{\dagger} \tag{3.52}$$

Then, we build implementation models in different noises for four different types of noises acting on qubits. Assuming that each qubit in quantum teleportation is in one of the noises, we first briefly describe the physical meanings of different noises and *Kraus*

1. Bit-flipping noise.

Bit-flipping noise can change the state of a quantum bit with probability P to flipping like $|0\rangle \rightarrow |1\rangle, |1\rangle \rightarrow |0\rangle$, and its *Kraus* operator is described as

$$E_1 = \sqrt{1-p}I, E_2 = \sqrt{p}\sigma_x. \tag{3.53}$$

2. Phase flip or phase decoherence noise.

Phase flip noise alters the phase of a qubit like $|1\rangle \rightarrow -|1\rangle$, and its *Kraus* operator is described as

$$E_1 = \sqrt{1-p}I, E_2 = \sqrt{p}\sigma_z \tag{3.54}$$

Phase flip noise is equivalent to phase decoherence, and the information loss of quantum state is considered in the noise process without energy loss. This is a typical decoherence model, because the phase flip noise destroys the quantum superposition of the qubits so that the off-diagonal elements of the density matrix reach zero.

3. Depolarization noise.

Depolarization noise is a very important noise. This type of noise requires a qubit and replaces it with probability P by mixed states $I/2$. It is considered as a "white noise." It will be represented as a density matrix of completely unpolarized states, i.e., $\langle\sigma_x\rangle = \langle\sigma_y\rangle = \langle\sigma_z\rangle = 0$. Its *Kraus* operator is described as

$$E_1 = \sqrt{1-3P/4}I, E_2 = \sqrt{P/4}\sigma_x \tag{3.55}$$

$$E_3 = \sqrt{P/4}\sigma_y, \quad E_4 = \sqrt{P/4}\sigma_z \tag{3.56}$$

4. Amplitude damping noise.

It is important to model the dissipation process of amplitude damping in several quantum systems, and its Kraus operator is described as

$$E_1 = \begin{bmatrix} 1 & 0 \\ 0 & \sqrt{1-P} \end{bmatrix}, = \begin{bmatrix} 0 & \sqrt{P} \\ 0 & 0 \end{bmatrix} \tag{3.57}$$

where P can be regarded as the decay probability from the excited state to the ground state in a two-level system, obtained $P = 1 - e^{-\frac{t}{T}}$ from a specific decoherence model (t represents time; T represents the specific time of the decoherence process).

3.2.1.3 Performance Analysis of Qubit Efficiency

Alice's Qubit Efficiency

We assume that the quantum channel is now immune to noise ($P_A = P_B = 0$) and the input state is in a quantum noise environment ($P_I \neq 0$). Each type of noise can be written as follows:

$$\langle \overline{F}_1 \rangle = \frac{2}{3}\left[1 - \frac{P_I}{2} + \frac{1-P_I}{2}\sin(2\theta)\sin(2\phi)\right] \tag{3.58}$$

$$\langle \overline{F}_2 \rangle = \frac{2}{3}\left[1 + \frac{1-2P_I}{2}\sin(2\theta)\sin(2\phi)\right] \tag{3.59}$$

$$\langle \overline{F}_3 \rangle = \frac{2}{3}\left[1 - \frac{P_I}{4} + \frac{1-P_I}{2}\sin(2\theta)\sin(2\phi)\right] \tag{3.60}$$

$$\langle \overline{F}_4 \rangle = \frac{2}{3}\left[1 - \frac{P_I}{4} + \frac{1}{2}\sqrt{1-P_I}\sin(2\theta)\sin(2\phi)\right] \tag{3.61}$$

The subscript in the left-hand side of Eq. (3.58) represents bit flip, the subscript in the left-hand side of Eq. (3.59) represents phase flip, the subscript in the left-hand side of Eq. (3.60) represents depolarizing noise, and the subscript in the left-hand side of Eq. (3.61) represents amplitude damping. When $\theta = \phi = \pm\frac{\pi}{4}, 1 - P_I > 0$, if $\sin(2\theta)\sin(2\phi) = 1$ could get the maximum value. In the expression of $\langle \overline{F}_2 \rangle$, $\sin(2\theta)\sin(2\phi)$ multiply $1 - 2P_I$, when $P_I < \frac{1}{2}$, the optimal setting is $\theta = \phi = \pm\frac{\pi}{4}$ and when $P_I > \frac{1}{2}$, the optimal setting is $\theta = -\phi = \pm\frac{\pi}{4}$. This means that if the input qubit phase flipped at a certain time $\left(P_I > \frac{1}{2}\right)$, Alice can change and improve the efficiency of quantum teleportation protocols ($\phi \to -\phi$). In addition, Alice can use the original measurement basis and either she or Bob can perform σ_z operations on the quantum channel, i.e., from $|B_1^\theta\rangle$ to $|B_1^{-\theta}\rangle$. We adopt the above optimal setting, and Eqs. (3.58) to (3.61) can be simplified as

$$\left\langle \overline{F}'_1 \right\rangle = 1 - \frac{2P_I}{3} \tag{3.62}$$

$$\left\langle \overline{F}'_2 \right\rangle = \frac{2}{3} + \frac{|1-2P_I|}{3} \tag{3.63}$$

$$\left\langle \overline{F}'_3 \right\rangle = 1 - \frac{P_I}{2} \tag{3.64}$$

$$\left\langle \overline{F}'_4 \right\rangle = \frac{2}{3} - \frac{P_I}{6} + \frac{1}{3}\sqrt{1-P_I} \tag{3.65}$$

As shown in Fig. 3.25, the noise of bit flip is the most serious of the four noises. When $0 \le P_I \le \frac{1}{2}$, the phase flip is as serious as the bit flip. On the other hand, from $P_I = 0$ to $P_I \approx 0.6$, amplitude damping is the most serious noise, followed by depolarization noise. And for the highest values of P_I, phase flip noise has the maximum average fidelity.

In order to more clearly show which qubits are affected by noise, the optimal efficiency of the protocol is written as $\left\langle \overline{F}'_{X,\phi,Y} \right\rangle$, where the first subscript X is the input state in the noise, the second subscript ϕ is the quantum bit of Alice on the quantum channel that is not affected by noise, and the third subscript Y is the quantum bit of Bob in the noise. The X and Y here could be any of the four kinds of noise.

When Bob's qubit is always affected by bit-flipping noise across four distinct noise environments, the corresponding optimal efficiency is described as follows:

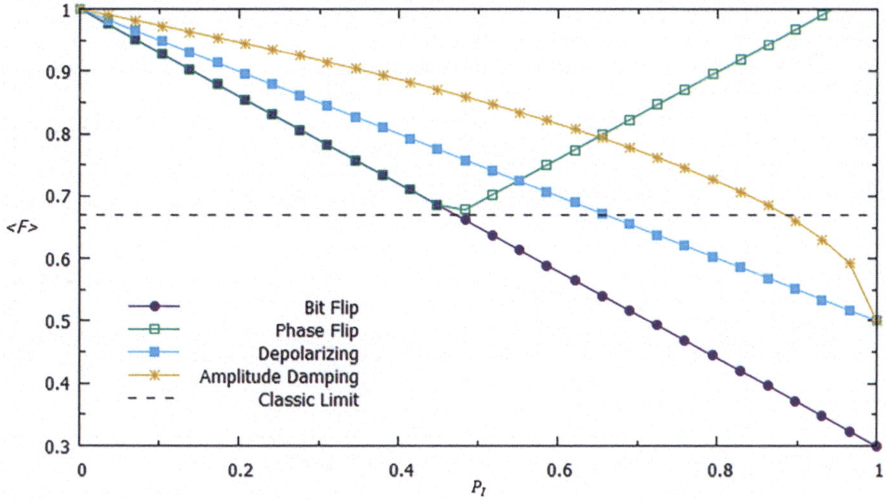

Fig. 3.25 The quantum teleportation efficiency of input state in a noisy channel

$$\left\langle \overline{F}'_{1,\phi,1} \right\rangle = 1 - \frac{2}{3}(P_I + P_B - 2P_I P_B) \tag{3.66}$$

$$\left\langle \overline{F}'_{1,\phi,2} \right\rangle = \frac{2}{3} - \frac{1}{3}[P_I - (1 - P_I)(1 - 2P_B)] \tag{3.67}$$

$$\left\langle \overline{F}'_{1,\phi,3} \right\rangle = 1 - \frac{P_B}{2} - \frac{2}{3}P_I(1 - P_B) \tag{3.68}$$

$$\left\langle \overline{F}'_{1,\phi,4} \right\rangle = \frac{2}{3} - \frac{1}{3}\left[P_I + \frac{P_B}{2}(1 - 2P_I)(1 - \cos(2\theta)) - (1 - P_I)\sqrt{1 - P_B}\sin(2\theta)\right]. \tag{3.69}$$

When $\theta = \phi = \pm\frac{\pi}{4}$, it is optimal parameters. When $P_B < \frac{1}{2}$, then $\theta = \phi = \pm\frac{\pi}{4}$ and when $P_B > \frac{1}{2}$, then $\theta = -\phi = \pm\frac{\pi}{4}$. When $\phi = \frac{\pi}{4}$, the given solution of θ is $d\left\langle \overline{F}'_{1,\phi,4} \right\rangle | d\theta$,

$$\tan(2\theta) = \frac{1(1 - P_I)\sqrt{1 - P_B}}{P_B(1 - 2P_I)} \tag{3.70}$$

When $P_B < \frac{1}{2}$, $\cos(2\theta) > 0$ and $\sin(2\theta) > 0$. And when $P_B > \frac{1}{2}$, $\cos(2\theta) < 0$.

This means that efficiency can be improved by reducing the amount of entanglement. When Bob's qubit is affected by amplitude damping noise, reducing the amount of entanglement can improve the efficiency (when the input state is pure). The same function is studied here when the input state is mixed. When $\theta \neq \frac{\pi}{4}$ and

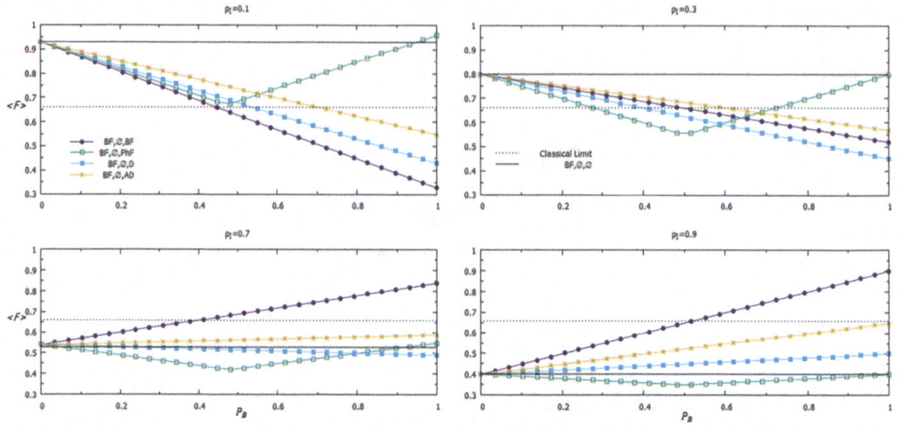

Fig. 3.26 The quantum teleportation efficiency of P_B and P_I in noisy channels

$P_B \neq 0$, less entanglement will lead to a better teleportation protocol as shown in Fig. 3.26.

In Fig. 3.26, the function of P_B has several values of P_I. It can be known that when $P_I > 0.5$, the stronger the noise, the higher the efficiency. In fact, when $P_I > 0.5$, you can see $\langle \overline{F}_1, \phi, \phi \rangle$ is lower than classical limits for P_B. However, by adding more noise into the protocol, Bob's qubit is flipped in the noisy environment, which improves the efficiency of the protocol when $P_I > 0.5$. For Bob's qubit, the efficiency of the protocol is always reduced in the bit-flipping noise when $P_I < 0.5$. It is worth mentioning that this happens when the input qubit is not affected by noise, but both two qubits in the quantum channel are affected by amplitude damping noise. When noise inevitably occurs, Bob can choose the noisy environment for quantum teleportation. In this case, the optimal noise depends on the unconventional values of P_I and P_B. For example, when $P_I < 0.3$, if Bob's qubit is subjected to amplitude damping noise, the protocol could implement better performance for the value of P_B not exceeding 0.6. However, when the value of P_B exceeds 0.6, we will get better results if Bob's qubit is subjected to phase flipping noise.

When the input qubit is affected by phase flipping noise and Bob's qubit is subjected to one of these noises, the optimal efficiency is described as follows:

$$\langle \overline{F}_2, \phi, 2 \rangle = \frac{2}{3}\left[1 + \frac{|(1-2P_I)(1-2P_B)|}{2}\right] \tag{3.71}$$

$$\langle \overline{F}_2, \phi, 1 \rangle = \frac{2}{3}\left[1 - \frac{P_B}{2} + \frac{|1-2P_I|(1-P_B)}{2}\right] \tag{3.72}$$

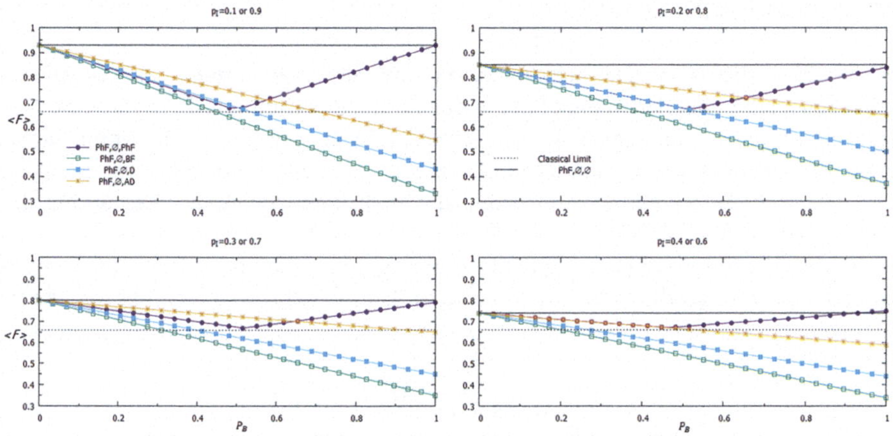

Fig. 3.27 The quantum teleportation efficiency of P_B and P_I in noisy channels

$$\langle \overline{F}_2, \phi, 3 \rangle = \frac{2}{3}\left[1 - \frac{P_B}{4} + \frac{|1-2P_I|(1-P_B)}{2}\right] \tag{3.73}$$

$$\langle \overline{F}_2, \phi, 4 \rangle = \frac{2}{3}\left[1 - \frac{P_B}{4} + \frac{P_B\cos(2\theta)}{4} + \frac{(1-2P_I)\sqrt{1-P_B}\sin(2\theta)}{2}\right] \tag{3.74}$$

When $(1 - 2P_I)(1 - 2P_B) > 0$ and $\theta = \phi = \pm\frac{\pi}{4}$, it will lead to optimal parameters in Eq. (3.71). When $(1 - 2P_I)(1 - 2P_B) < 0$, then $\theta = -\phi = \pm\frac{\pi}{4}$. In Eqs. (3.70) and (3.71), if $(1 - 2P_I) > 0$, $\theta = \phi = \pm\frac{\pi}{4}$ and when $(1 - 2P_I) < 0$, $\theta = -\phi = \pm\frac{\pi}{4}$. In Eq. (3.72), when $\phi = \frac{\pi}{4}$, the given solution of θ is $d\left\langle \overline{F}'_{1,\phi,4} \right\rangle | d\theta$:

$$\tan(2\theta) = \frac{2(1-2P_I)\sqrt{1-P_B}}{P_B} \tag{3.75}$$

When $P_I < \frac{1}{2}$, $\cos(2\theta) > 0$ and $\sin(2\theta) > 0$. And when $P_I > \frac{1}{2}$, $\sin(2\theta) < 0$. When $P_B \neq 0$, minimum entanglement can achieve maximum efficiency.

In Fig. 3.27, P_B as a function has corresponding value of P_I. On the contrary, for the input qubit in bit-flipping noise, it's equivalent to no noise ($P_B = 0$). If Bob's qubit is put into the noise when $P_B \neq 0$, the noise can be increased without improving the efficiency. However, if noise is unavoidable and Bob can choose different noisy channels, he can improve the efficiency of the protocol by choosing the right noise. His choice depends on the probability of P_B that is acting on the qubit in a noisy channel. At the same time, his qubit in a noisy environment is shown in Fig. 3.27.

It can be seen that if Bob's qubit is subjected to amplitude damping noise, P_B can achieve maximum efficiency from 0 to 0.6. On the other hand, it works better when the input quantum state and Bob's quantum state are subjected to the same type of noise, namely, the value of P_B from 0.6 to 1.

All the Qubits in the Noise

When all of Alice's particles are subjected to the same type of noise, such as Alice's input state and entangled channel in the quantum teleportation task. In this case, the input qubit and the entangled state of Alice are in the same noise environment, so they are subjected to the same type of noise in the same time span, which means $P_I = P_B = P$. Bob's particles are subjected to different types of noise during the travel from Alice to Bob.

If Alice's qubit is subjected to the bit-flipping noise, the optimal efficiency could be obtained through the influence of noise on Bob's qubit, which can be written as follows:

$$\langle \overline{F}_1, 1, \phi \rangle = 1 - \frac{4P(1-P)}{3} \tag{3.76}$$

$$\langle \overline{F}_1, 1, 1 \rangle = 1 - \frac{2P_B}{3} - \frac{4P(1-P)(1-2P_B)}{3} \tag{3.77}$$

$$\langle \overline{F}_1, 1, 2 \rangle = \frac{2}{3} - \frac{2P(1-P)}{3} + \frac{[1-2P(1-P)]\,|\,1-2P_B}{3} \tag{3.78}$$

$$\langle \overline{F}_1, 1, 3 \rangle = 1 - \frac{P_B}{2} - \frac{4P(1-P)(1-P_B)}{3} \tag{3.79}$$

$$\begin{aligned} \overline{F}_1, 1, 4 \rangle = \frac{2}{3} - \frac{P_B}{6} - \frac{2P(1-P)(1-P_B)}{3} + \frac{(1-2P)^2 P_B \cos(2\theta)}{6} \\ + \frac{[1-2P(1-P)]\sqrt{1-P_B}\sin(2\theta)}{3} \end{aligned}. \tag{3.80}$$

The optimal parameters given in Eqs. (3.76), (3.77), and (3.78) are $\theta = \phi = \pm\frac{\pi}{4}$. In Eq. (3.77), if $P_B < \frac{1}{2}$, then $\theta = \phi = \pm\frac{\pi}{4}$ and if $P_B > \frac{1}{2}$, then $\theta = -\phi = \pm\frac{\pi}{4}$. In Eq. (3.79), when one of the optimal parameters is $\phi = \frac{\pi}{4}$and the given solution of θ is

$$\tan(2\theta) = \frac{2[1-2P(1-P)]\sqrt{1-P_B}}{(1-2P)^2 P_B}. \tag{3.81}$$

In Eq. (3.79), $\cos(2\theta) > 0$ and $\sin(2\theta) > 0$.

In Fig. 3.28, P_B as a function has corresponding value of P_I.

It can be seen that adding more noise does not improve the efficiency of the protocol. On the contrary, only the input bits are affected by the bit-flipping noise, and Bob's qubits are also affected by the bit-flipping noise. In Fig. 3.28, if noise is unavoidable, Bob selects the correct noisy channel according to the value of P_B and P_I to improve the efficiency of the protocol.

In fact, in these three noisy cases, Bob's noisy channel is different from Alice's noisy channel, which can improve the efficiency of the protocol. In particular, if Bob's qubit is subjected to amplitude damping noise, the best performance is achieved with the lowest value of P_B. For maximum values of P_B, the phase flip

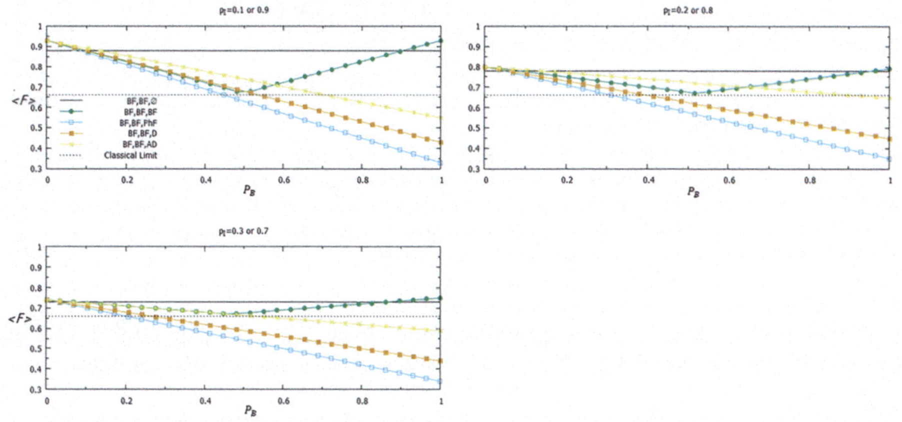

Fig. 3.28 The quantum teleportation efficiency when Alice's qubits are subjected to bit-flipping noise

channel is a layout that provides the best performance. Finally, when at least one particle acts on the amplitude damped channel, the optimal entanglement trend of the quantum channel is obtained, which means the less the entanglement, the better the performance.

If Alice is affected by the other three kinds of noise, it is not found that any noise improves the overall efficiency. In these three noise cases, Alice's noise channel is different from Bob's noise channel, which can improve the efficiency of the protocol. If Bob's qubit receives amplitude damping noise, the lowest value of P_B achieves the best performance. For the highest value of P_B, the phase reversal channel is a layout that provides the best performance. When at least one particle acts on the amplitude damped channel, the optimal entanglement trend of the quantum channel is obtained, which means the less the entanglement, the better the performance.

Qubits on Quantum Channel

In the process of quantum teleportation, the channel particles may be affected by noise. In this case, the qubits during Alice and Bob's communication are in the same noisy environment, and they are subjected to the same noise at the same time, which means $P_A = P_B = P$.

If the channel particle suffers from bit-flipping noise, we obtain the following optimal efficiency according to the noise type of the qubit:

$$\langle \overline{F}\phi, 1, 1 \rangle = 1 - \frac{4P(1-P)}{3} \tag{3.82}$$

$$\langle \overline{F}_1, 1, 1 \rangle = 1 - \frac{2P_I}{3} - \frac{4(1-2P_I)P(1-P)}{3} \tag{3.83}$$

$$\langle \overline{F}_2, 1, 1 \rangle = \frac{2}{3} - \frac{2P(1-P)}{3} + \frac{|1-2P_I|[1-2P(1-P)]}{3} \tag{3.84}$$

$$\langle \overline{F}_3, 1, 1 \rangle = 1 - \frac{P_I}{2} - \frac{4(1-P_I)P(1-P)}{3} \tag{3.85}$$

$$\langle \overline{F}_4, 1, 1 \rangle = \frac{2}{3} - \frac{P_I}{6} - \frac{2(1-P_I)P(1-P)}{3} + \frac{\sqrt{1-P_I}[1-2P(1-P)]}{3} \tag{3.86}$$

When $\theta = \phi = \pm \frac{\pi}{4}$, then the optimal parameters are given by Eqs. (3.83), (3.84), and (3.85). And in Eq. (3.86), if $P_I < \frac{1}{2}$, then $\theta = \phi = \pm \frac{\pi}{4}$ and if $P_I > \frac{1}{2}$, then $\theta = -\phi = \pm \frac{\pi}{4}$.

When $\langle \overline{F}_1, 1, X \rangle = \langle \overline{F}_X, 1, 1 \rangle$ and $X = \phi$, 1, 2, 3, 4. If $P_B = P_I$ and $= \frac{\pi}{4}$, 2e have analyzed and discussed this situation before.

3.2.2 *Collective Noise Immunity Model Based on DFS*

In this section, based on the quantum entanglement evolution model, the immune noise model is established by using non-decoherent subspace for joint noise.

3.2.2.1 Build the Immune Model of Collective Dephasing Noise

According to the influence of the collective dephasing noise on the quantum state, the logical qubit we constructed that can be immune to the dephasing noise is $|0_{dp}\rangle = |01\rangle \equiv |0\rangle|1\rangle$ and $|1_{dp}\rangle = |10\rangle \equiv |1\rangle|0\rangle$ (the subscript "*dp*" represents the logical state that can resist noise). Any superposition of $|0_{dp}\rangle$ and $|1_{dp}\rangle$ can form a decoherence-free subspace (DFS) which can be immune to collective dephasing noise. In this space, another measurement basis is $|\pm_{dp}\rangle = \frac{1}{\sqrt{2}}(|0_{dp}\rangle \pm |1_{dp}\rangle) \equiv \frac{1}{\sqrt{2}}(|0\rangle|1\rangle \pm |1\rangle|0\rangle)$, which is also immune to collective noise.

The two non-orthogonal measurement bases under the subspace (DFS) are $\{|0_{dp}\rangle, ||_{dp}\rangle\}$ and $\left\{|+\rangle_{dp} = \frac{1}{\sqrt{2}}(|0_{dp}\rangle + |1_{dp}\rangle), |-\rangle_{dp} = \frac{1}{\sqrt{2}}(|0_{dp}\rangle - |1_{dp}\rangle)\right\}$.

Taking the four-qubit cluster state as an example, the four-qubit cluster state can be expressed as

$$\begin{aligned} |\varphi^+\rangle_{XYZM} = & \frac{1}{2}(|0000\rangle + |0011\rangle + |1100\rangle + |1111\rangle)_{XYZM} \\ & = \frac{1}{\sqrt{2}}(|00\rangle + |11\rangle)_{XY} \bigotimes \frac{1}{\sqrt{2}}(|00\rangle + |11\rangle)_{ZM} \end{aligned} . \tag{3.87}$$

Suppose the sender Alice has the particles of cluster state (*X*, *Y*, *Z*, and *M*) that are in maximum entanglement. If she wants to send particles *Y* and *M* to the receiver Bob. In order to prevent the influence of dephasing noise, Alice first turns particles *Y* and *M* into logical quantum states, namely, $|0\rangle_Y \equiv |0\rangle_{Y1}|1\rangle_{Y2}, |1\rangle_Y \equiv |1\rangle_{Y1}|0\rangle_{Y2}$ and $|0\rangle_M \equiv |0\rangle_{M1}|1\rangle_{M2}, |1\rangle_M \equiv |1\rangle_{M1}|0\rangle_{M2}$. $0\rangle_{M1}|1\rangle_{M2}, |1\rangle_M \equiv |1\rangle_{M1}|0\rangle_{M2}$. Then $|\varphi^+\rangle_{XYZM} \equiv \left|\varphi^+_{dp}\right\rangle_{XYZM} = \left|\varphi^+_{dp}\right\rangle_{XY} \otimes \left|\varphi^+_{dp}\right\rangle_{ZM} = \frac{1}{\sqrt{2}}(|0\rangle_X|01\rangle_{Y_1Y_2} + |1\rangle_X|10\rangle_{Y_1Y_2}) \otimes \frac{1}{\sqrt{2}}(|0\rangle_Z|01\rangle_{M_1M_2} + |1\rangle_Z|10\rangle_{M_1M_2})$.

Since the phase of the photon is constant as it travels through the channel, the mode changes, namely, $|1\rangle \to |1\rangle$. In $e^{i\theta}$, θ fluctuates with time. When $\theta = 0$, the phase and mode are constant. The sender Alice only needs to prepare two-dimensional Bell state $|\varphi^+\rangle_{XYZM}$ of 4 qubits. Then the particles Y_1Y_2 and M_1M_2 are sent to the receiver Bob. After receiving the logical quantum state, Bob conducts appropriate transformation, and then we get the original information. Here, Bob needs to conduct controlled-NOT (CNOT) operation, taking Y_1 and M_1 as controlled qubits and Y_2 and M_2 as target qubits, so we can get $\frac{1}{\sqrt{2}}(0\rangle_X|0\rangle_{Y_1}|1\rangle_{Y_2} + |1\rangle_X|1\rangle_{Y_1}|1\rangle_{Y_2}) \otimes \frac{1}{\sqrt{2}}(0\rangle_Z|0\rangle_{M_1}|1\rangle_{M_2} + |1\rangle_Z|1\rangle_{M_1}|1\rangle_{M_2})$, that is, $\frac{1}{\sqrt{2}}\{(|0\rangle_X|0\rangle_{Y_1} + |1\rangle_X|1\rangle_{Y_1})|1\rangle_{Y_2}\} \otimes \frac{1}{\sqrt{2}}\{(|0\rangle_Z|0\rangle_{M_1} + |1\rangle_Z|1\rangle_{M_1})|1\rangle_{M_2}\}$.

It can be seen that particles Y_1 and M_1 owned by Bob and particles *X* and *Z* owned by Alice are still in the maximum entangled state $|\varphi^+\rangle$. Therefore, it can be understood that particles *Y* and *M* are immune to collective dephasing noise during transmission. In other words, they are not affected by collective dephasing noise.

When the photon is transmitted in the channel $|1\rangle \to |1\rangle$, in $e^{i\theta}$, when $\theta \neq 0$, the phase is constant, but the mode changes. In $\left|\varphi^+_{dp}\right\rangle$ under the influence of different θ_1 and θ_1, the four-qubit cluster state is taken as an example to establish a quantum model immune to dephasing noise:

$$
\begin{aligned}
\left|\varphi^+_{dp}\right._{1234} &= \frac{1}{\sqrt{2}}(|0_{dp}|0_{dp} + |1_{dp}1_{dp})_{12} \otimes \frac{1}{\sqrt{2}}(|0_{dp}|0_{dp} + |1_{dp}1_{dp})_{34} \\
&= \frac{1}{\sqrt{2}}[(e^{i\theta_1}|01e^{i\theta_2}|01) + (e^{i\theta_1}|10e^{i\theta_2}|10)]_{12} \otimes \frac{1}{\sqrt{2}}[(e^{i\theta_1}|01e^{i\theta_2}|01) + (e^{i\theta_1}|10e^{i\theta_2}|10)]_{34} \\
&= \frac{1}{\sqrt{2}}[(e^{i\theta_1+i\theta_2}|01|01) + (e^{i\theta_1+i\theta_2}|10|10)]_{12} \otimes \frac{1}{\sqrt{2}}[(e^{i\theta_1+i\theta_2}|01|01) + (e^{i\theta_1+i\theta_2}|10|10)]_{34} \\
&= \frac{e^{i\theta_1+i\theta_2}}{\sqrt{2}}[(|01\rangle|01\rangle + |10\rangle|10\rangle)]_{12} \otimes \frac{e^{i\theta_1+i\theta_2}}{\sqrt{2}}[(|01\rangle|01\rangle + |10\rangle|10\rangle)]_{34} \\
&= \frac{e^{i\theta_1+i\theta_2}}{\sqrt{2}}[(|00\rangle|00\rangle + |11\rangle|11\rangle)]_{12} \otimes \frac{e^{i\theta_1+i\theta_2}}{\sqrt{2}}[(|00\rangle|00\rangle + |11\rangle|11\rangle)]_{34} \\
&= e^{i\theta_1+i\theta_2}\left(\left|\varphi^+_{dp}\right._{12} \otimes \left|\varphi^+_{dp}\right._{34}\right)
\end{aligned}
\tag{3.88}
$$

According to the above equation, when θ fluctuates with time under the collective dephasing noise, particles 2 and 4 owned by Bob and particles 1 and 3 owned by

Alice are still in the maximum entangled state $|\varphi^{+}\rangle$. Thus, it can be understood that particles 2 and 4 are immune to collective dephasing noise during transmission. In other words, they are not affected by collective dephasing noise.

3.2.2.2 Build an Immune Model of Collective-Rotation Noise

We express the two photons in the form of logical quantum states: $|0_r\rangle \equiv |\Phi^{+}\rangle = \frac{1}{\sqrt{2}}(|00\rangle + |11\rangle)$ and $|1_r\rangle \equiv |\Psi^{-}\rangle = \frac{1}{\sqrt{2}}(|01\rangle - |10\rangle)$ (the subscript r represents the logical state that can resist collective noise, and $|\Phi^{+}\rangle$ and $|\Psi^{-}\rangle$ are just two states of the four original Bell states that they don't change their states under rotational noise). Superposition of two logical states $|\Phi^{+}\rangle$ and $|\Psi^{-}\rangle$ can form a decoherence-free subspace (DFS) which can be immune to noise. In this space, another measurement bases are $\{|0_{dp}\rangle, \|_{dp}\rangle\}$, $|+_r\rangle = \frac{1}{\sqrt{2}}(|0_r\rangle + |1_r\rangle) = \frac{1}{\sqrt{2}}(|\Phi^{+}\rangle + |\Psi^{-}\rangle)$, and $|-_r\rangle = \frac{1}{\sqrt{2}}(|0_r\rangle - \quad |1_r\rangle) = \frac{1}{\sqrt{2}}(|\Phi^{+}\rangle - |\Psi^{-}\rangle)$, which are also immune to collective noise.

The two non-orthogonal measurement bases under the subspace (DFS) are $\{|0_r\rangle, |1_r\rangle\}$ and $\left\{|+\rangle_r = \frac{1}{\sqrt{2}}(|0_r\rangle + |1_r\rangle), |-\rangle_r = \frac{1}{\sqrt{2}}(|0_r\rangle - |1_r\rangle)\right\}$.

In quantum teleportation for communication, the sender first turns the physical photons into logical quantum state and then sends them in channel. The receiver receives the logical quantum states and the original quantum state can be restored by taking appropriate unitary operations. Therefore, it can be guaranteed that photons are not affected by noise in transmission process. The accuracy of photon transmission is guaranteed.

Taking the four-qubit cluster state as an example, the four-qubit cluster state can be expressed as

$$\begin{aligned} |\varphi^{+}\rangle_{XYZM} &= \frac{1}{2}(|0000\rangle + |0011\rangle + |1100\rangle + |1111\rangle)_{XYZM} \\ &= \frac{1}{\sqrt{2}}(|00\rangle + |11\rangle)_{XY} \bigotimes \frac{1}{\sqrt{2}}(|00\rangle + |11\rangle)_{ZM} \end{aligned}. \tag{3.89}$$

Suppose the sender Alice has the particles of cluster state (X, Y, Z, and M) that are in maximum entanglement. If she wants to send particles Y and M to the receiver Bob. In order to prevent the influence of rotational noise, Alice first turns particles Y and particle M into logical quantum states, namely, $|0\rangle_Y = |\Phi^{+}\rangle_{Y_1Y_2}$, $|1\rangle_Y = |\Psi^{+}\rangle_{Y_1Y_2}$ 和 $|0\rangle_M = |\Phi^{+}\rangle_{M_1M_2}$, $|1\rangle_M = |\Psi^{+}\rangle_{M_1M_2}$. Therefore, $|\varphi^{+}_{XZM}\rangle$ can be expressed as

$$|\varphi_r^+\rangle_{XYZM} = \frac{1}{\sqrt{2}}(0\rangle_X|\Phi^+\rangle_{Y_1Y_2}+|1\rangle_X|\Psi^-\rangle_{Y_1Y_2})\bigotimes\frac{1}{\sqrt{2}}$$
$$\times\left(|0\rangle_z|\Phi^+\rangle_{M_1M_2}+|1\rangle_Z|\Psi^-\rangle_{M_1M_2}\right)=\frac{1}{2}[0\rangle_X(|00\rangle_{Y_1Y_2}+|11\rangle_{Y_1Y_2})+$$
$$|1\rangle_X(|01\rangle_{Y_1Y_2}-|10\rangle_{Y_1Y_2})\bigotimes\bigotimes\frac{1}{2}$$
$$\times[0\rangle_Z(|00\rangle_{M_1M_2}+|11\rangle_{M_1M_2})+|1\rangle_Z(|01\rangle_{M_1M_2}-|10\rangle_{M_1M_2})].$$

Therefore, the sender Alice only needs to prepare two-dimensional Bell state of 4 qubits: $|\varphi^+\rangle_{XYZM}$. The particles Y_1Y_2 and M_1M_2 are then sent to the receiver Bob. Alice and Bob performed swap gate operations on particles X, Y_1Y_2 and Z, M_1M_2, respectively, namely, $S_X\bigotimes S_{Y_1}\bigotimes S_{Y_2}$ and $S_Z\bigotimes S_{M_1}\bigotimes S_{M_2}$ (in swap gate, $S=\begin{pmatrix}1&0\\0&i\end{pmatrix}=\alpha|0\rangle+i\beta|1\rangle$, and α and β are complex numbers satisfying $|\alpha|^2+|\beta|^2=1$). Then they performed Hadamard gate operation, namely, $H_X\bigotimes H_{Y_1}\bigotimes H_{Y_2}$ and $H_Z\bigotimes H_{M_1}\bigotimes H_{M_2}$ (Hadamard gate: $H=\frac{1}{\sqrt{2}}\begin{pmatrix}1&1\\1&-1\end{pmatrix}$, $|0\rangle\equiv\frac{(|0\rangle+|1\rangle)}{\sqrt{2}}$, $|1\rangle\equiv\frac{(|0\rangle-|1\rangle)}{\sqrt{2}}$).

At the same time, the joint quantum state of particles X, Y_1Y_2 and Z, M_1M_2 will become

$$|\wp_r\rangle_{XYZM}=\frac{1}{\sqrt{2}}(|1\rangle_X|10\rangle_{Y_1Y_2}+|0\rangle_X|01\rangle_{Y_1Y_2})\bigotimes\frac{1}{\sqrt{2}}(|1\rangle_Z|10\rangle_{M_1M_2}+|0\rangle_Z|01\rangle_{M_1M_2}).\tag{3.90}$$

After Bob received the logical quantum state, he performed the appropriate operation and then we can get the original information. Here, Bob needs to perform CNOT operation (CNOT gate $=\begin{pmatrix}I&0\\0&I\end{pmatrix}$, $I=\begin{pmatrix}1&0\\0&1\end{pmatrix}=\alpha|0\rangle+\beta|1\rangle$, and α and β are complex numbers satisfying $|\alpha|^2+|\beta|^2=1$) with Y_1 and M_1 as controlled qubits and Y_2 and M_2 as target qubits; then we can get,

$$\begin{aligned}|\wp_r\rangle'_{XYZM}&=\frac{1}{\sqrt{2}}(|1\rangle_X|11\rangle_{Y_1Y_2}+|0\rangle_X|01\rangle_{Y_1Y_2})\bigotimes\frac{1}{\sqrt{2}}(|1\rangle_Z|11\rangle_{M_1M_2}+|0\rangle_Z|01\rangle_{M_1M_2})\\&=\frac{1}{\sqrt{2}}[(|0\rangle_X|0\rangle_{Y_1}+|1\rangle_X|1\rangle_{Y_1})|1\rangle_{Y2}]\bigotimes\frac{1}{\sqrt{2}}[(|0\rangle_Z|0\rangle_{M_1}+|1\rangle_Z|1\rangle_{M_1})|1\rangle_{M_2}]\end{aligned}$$

it can be seen that particles Y_1 and M_1 owned by Bob and particles X and Z owned by Alice are still in the maximum entangled state $|\varphi^+{}_{XYZM}\rangle$. Therefore, it can be understood that particles Y and M are immune to collective-rotation noise during transmission. In other words, they are not affected by collective-rotation noise.

Similarly, for that the phase of the photon is constant when it travels through the channel; the mode will change in the collective-rotation noise, namely, in $e^{i\theta}$, where

θ fluctuates with time. In $\left|\varphi_{dp}^{+}\right\rangle$ under the influence of different θ_1 and θ_1, the four-qubit cluster state is taken as an example to establish a quantum model immune to rotational noise (the quantum model of tolerance can be expressed by linear formula):

$$
\begin{aligned}
|\Phi_r^+\rangle_{1234} &= \frac{1}{\sqrt{2}}(||0_r\rangle|0_r\rangle + |1_r\rangle|1_r\rangle)_{12} \bigotimes \frac{1}{\sqrt{2}}(|0_r\rangle|0_r\rangle + |1_r\rangle|1_r\rangle)_{34} \\
&= \frac{1}{\sqrt{2}}(|\Phi^+\rangle|\Phi^+\rangle + |\Psi^-\rangle|\Psi^-\rangle)_{12} \bigotimes \frac{1}{\sqrt{2}}(|\Phi^+\rangle|\Phi^+\rangle + |\Psi^-\rangle|\Psi^-\rangle)_{34} \\
&= \frac{1}{2\sqrt{2}}(|0_r\rangle|0_r\rangle + |1_r\rangle|1_r\rangle) \bigotimes (|0_r\rangle|0_r\rangle + |1_r\rangle|1_r\rangle) \\
&+ (|0_r\rangle|0_r\rangle - |1_r\rangle|1_r\rangle) \bigotimes (|0_r\rangle|0_r\rangle - ||1_r\rangle|1_r\rangle]. \\
&= \frac{1}{2\sqrt{2}}[(|00\rangle + |11\rangle)(|00\rangle + |11\rangle) + (|01\rangle - |10\rangle)(|01\rangle - |10\rangle)]_{12} \\
&\bigotimes \frac{1}{2\sqrt{2}}[(|00\rangle + |11\rangle)(|00\rangle + |11\rangle) + (|01\rangle - |10\rangle)(|01\rangle - |10\rangle)]_{34} \\
&= \frac{1}{\sqrt{2}}(|\Phi^+\rangle|\Phi^+\rangle + |\Psi^-\rangle|\Psi^-\rangle)_{12} \bigotimes \frac{1}{\sqrt{2}}(|\Phi^+\rangle|\Phi^+\rangle + |\Psi^-\rangle|\Psi^-\rangle \partial_{34}
\end{aligned}
\tag{3.91}
$$

According to the above equation, when θ fluctuates with time, particles 2 and 4 owned by Bob and particles 1 and 3 owned by Alice are still in the maximum entangled state $|\varphi^+\rangle$. Thus, it can be understood that particles 2 and 4 are immune to the collective-rotation noise during transmission. In other words, they are not affected by collective-rotation noise.

3.2.2.3 Immune Collective Noise Model of Two-Dimensional Four-Qubit Hyperentangled State

In the dephasing noise channel, we use the two-dimensional four-qubit hyperentangled state under the polarization and space degrees of freedom $|\Phi^+{}_{XYZM}\rangle = \frac{1}{2}(|0000\rangle + |0011\rangle + | \quad |1100\rangle + 1111\rangle)_{XYZM} \bigotimes (|a_1b_1\rangle \ + |a_2b_2\rangle)_{XY} \times \bigotimes (|a_1b_1\rangle \ + |a_2b_2\rangle)_{ZM}$ as an example. If Alice wants to transmit particles Y and M in the collective dephasing noise channel, the logical qubit immune to dephasing noise must be constructed: $|0_Y\rangle = |01\rangle \equiv |0\rangle_{Y_1} |1\rangle_{Y_2}$, $|0_Z\rangle = |01\rangle \equiv |0\rangle_{M_1} |\rangle_{M_2}$, $|1_Y\rangle = |10\rangle \equiv |1\rangle_{Y_1} |0\rangle_{Y_2}$, $|1_Z\rangle = |10\rangle \equiv |1\rangle_{M_1} |0\rangle_{M_2}$. Then, the two-dimensional four-qubit hyperentangled state can be expressed as

$$\begin{aligned}|\Phi_{XYZM}^{+}\rangle_{PS} = & \frac{1}{\sqrt{2}}\left(|0\rangle_X + |01\rangle_{Y_1Y_2} + |1\rangle_X + |10\rangle_{Y_{Y_2}}\right)_{XY} \\ & \bigotimes \frac{1}{\sqrt{2}}(|0\rangle_Z + |01\rangle_{M_1M_2} + |1\rangle_Z + |10\rangle_{M_1M_2})_{ZM} \\ & \bigotimes (|a_1b_1c_1\rangle + |a_2b_2c_2\rangle)_{XY} \bigotimes (|a_1b_1c_1\rangle + |a_2b_2c_2\rangle)_{ZM}\end{aligned} \quad . \tag{3.92}$$

Alice sent particles Y_1Y_2 and M_1M_2 to the receiver Bob. When Bob received the logical qubit, he performed CNOT operation with Y_1 and M_1 as controlled qubits and Y_2 and M_2 as target qubits; then we can get the new state

$$\begin{aligned}|\Phi_{XYZM}^{+}\rangle_{PS}' = & \frac{1}{\sqrt{2}}[(|0\rangle_X|0\rangle_{Y_1} + |1\rangle_X|1\rangle_{Y_1})|1\rangle_{Y_2}] \bigotimes \frac{1}{\sqrt{2}}[(|0\rangle_Z|0\rangle_{M_1} + |1\rangle_Z|1\rangle_{M_1})|1\rangle_{M_2}] \\ & \bigotimes (|a_1b_1c_1\rangle + |a_2b_2c_2\rangle)_{XYY} \bigotimes (|a_1b_1c_1\rangle + |a_2b_2c_2\rangle)_{ZM_1M_2}\end{aligned} \quad . \tag{3.93}$$

It can be seen from $|\Phi_{XYZM}^{+}\rangle_{PS}$ particles Y_1 and M_1 owned by Bob and particles X and Z owned by Alice are both in the maximum entangled state $|\Phi_{XYZM}^{+}\rangle$. Therefore, it can be understood that particles Y and M are immune to collective dephasing noise during transmission. In other words, they are not affected by collective dephasing noise.

In the rotational noise channel, we use the two-dimensional four-qubit hyperentangled state under the polarization and space degrees of freedom $|\Phi_{XYZM}{}^{+}\rangle = \frac{1}{2}(|0000\rangle + |0011\rangle + |1100\rangle + |1111\rangle)_{XYZM} \times \bigotimes (|a_1b_1\rangle + |a_2b_2\rangle)_{XY} \bigotimes (|a_1b_1\rangle + |a_2b_2\rangle)_{ZM}$ as an example. If Alice wants to transmit particles Y and M in the collective-rotation noise channel, the logical qubit immune to dephasing noise also must be constructed: $|0_Y\rangle = |\Phi^{+}\rangle = \frac{1}{\sqrt{2}}(00\rangle + |11\rangle)_{Y_1Y_2}$, $|1_Y\rangle = |\Psi^{+}\rangle = \frac{1}{\sqrt{2}}(01\rangle - |10\rangle)_{Y_1Y_2}$, $|0_M\rangle = |\Phi^{+}\rangle = \frac{1}{\sqrt{2}}(|00\rangle + |11\rangle)_{M_1M_2}$, $|1_M\rangle = |\Psi^{+}\rangle = \frac{1}{\sqrt{2}}(|01\rangle - |10\rangle)_{M_1M_2}$. Then, the two-dimensional four-qubit hyperentangled state can be expressed as

$$\begin{aligned}|\Phi_{XZZM}^{+}\rangle_{PS}^{r} = & \frac{1}{\sqrt{2}}\left(|0\rangle_X + |\Phi^{+}\rangle_{Y_1Y_2} + |1\rangle_X + |\Psi^{-}\rangle_{YY_1Y_X}\right)_{XY} \\ & \otimes \frac{1}{\sqrt{2}}\left(|0\rangle_Z + |\Phi^{+}\rangle_{M_1M_2} + |1\rangle_Z + |\Psi^{-}\rangle_{M_1M_2}\right)_{ZM} \\ & \otimes(|a_1b_1c_1\rangle + |a_2b_2c_2\rangle)_{XY_1Y_2} \otimes (|a_1b_1c_1\rangle + |a_2b_2c_2\rangle)_{ZM_1M_2} \\ = & \frac{1}{\sqrt{2}}[|0\rangle_X(|00\rangle_{Y_1Y_2} + |11\rangle_{Y_1Y_2}) + |1\rangle_X(|01\rangle_{Y_1Y_2} - |10\rangle_{Y_1Y_2})] \otimes (|a_1b_1c_1\rangle + |a_2b_2c_2\rangle)_{XY_1Y_2} \otimes \\ & \frac{1}{\sqrt{2}}[|0\rangle_Z(|00\rangle_{M_1M_2} + |11\rangle_{M_1M_2}) + |1\rangle_Z(|01\rangle_{M_1M_2} - |10\rangle_{M_1M_2})] \otimes (|a_1b_1c_1\rangle + |a_2b_2c_2\rangle)_{ZM_1M_2}\end{aligned} \quad . \tag{3.94}$$

Alice sent particles Y_1Y_2 and M_1M_2 to the receiver Bob. Alice and Bob performed swap gate operations on particles X, Y_1Y_2 and Z, M_1M_2, respectively, namely, $S_X \bigotimes S_{Y_1} \bigotimes S_{Y_2}$ and $S_Z \bigotimes S_{M_1} \bigotimes S_{M_2}$.

At the same time, the joint quantum state of particles X, Y_1Y_2 and Z, M_1M_2 will become

$$|\Phi^+_{XYZM}\rangle^{r\,\prime}_{PS} = \frac{1}{\sqrt{2}}(|1\rangle_X|10\rangle_{Y_1Y_2} + |0\rangle_X|01\rangle_{Y_1Y_2}) \bigotimes (|a_1b_1c_1\rangle + |a_2b_2c_2\rangle)_{XY_1Y_2}. \tag{3.95}$$

Similarly, when Bob received the logical qubit, he performed CNOT operation with Y_1 and M_1 as controlled qubits and Y_2 and M_2 as target qubits; then we can get the new state

$$|\Phi^+_{XYZM}\rangle^{r\,\prime\prime}_{PS} = \frac{1}{\sqrt{2}}[(|0\rangle_X|0\rangle_{Y_1} + |1\rangle_X|1\rangle_{Y_1})|1\rangle_{Y_2}] \bigotimes (|a_1b_1c_1\rangle + |a_2b_2c_2\rangle)_{XY_1Y_2} \bigotimes$$
$$\frac{1}{\sqrt{2}}[(|0\rangle_Z|0\rangle_{M_1} + |1\rangle_Z|1\rangle_{M_1})|1\rangle_{M_2}] \bigotimes (|a_1b_1c_1\rangle + |a_2b_2c_2\rangle)_{ZM_1M_2} \tag{3.96}$$

According to the state of $|\Phi^+_{XYZM}\rangle^{r\,\prime\prime}_{PS}$, particles Y_1 and M_1 owned by Bob and particles X and Z owned by Alice are both in the maximum hyperentangled state $|\Phi^+_{XYZM}\rangle$. Thus, it can be understood that particles Y and M are immune to the dephasing noise during transmission. In other words, they are not affected by dephasing noise.

3.3 Channel Capacity Coding

The channel capacity problem based on noise seriously affects the quality of quantum teleportation, and quantum teleportation is always under single degree of freedom at present. The reason for this is that in many application scenarios, people often do not need to locate the quantum with multiple degrees of freedom but only operate in one state. How to prevent low channel capacity under single degree of freedom becomes a key problem if the quantum is not encoded under multiple degrees of freedom. Our research found that a feasible solution is to design more powerful quantum codes. Quantum graph states can be used for quantum coding in order to obtain more perfect theoretical results. We design a method of graph state basis of quantum graph coding on tree graph and forest graph under multiple degrees of freedom. The quantum coherence under different noise channels is obtained by analyzing the information of coherence and channel capacity. And we effectively calculate the noise volume capacity of approximation, computing speed and noise margin of channel-transmitted quantum information. We then obtained the area where the quantum information can be transmitted under noise channels.

3.3.1 Mathematical Description of Quantum Concatenated Code

The mathematical descriptions of graph state basis, tree graph, and forest graph are the basis of the study of quantum graph state coding. Firstly, based on the study of the entanglement evolutionary immune model, a mathematical model is established to describe the input and output states of the graph state basis to simplify von Neumann entropy and obtain decoherence information easily. Secondly, different from the channel coding of the traditional coding technology, the mathematical descriptions of tree graph and forest graph based on graph state basis in quantum graph state coding need to determine the coding indexes and key parameters of graph state basis.

We assume that the internal encoding and external encoding are used to construct the quantum concatenated code. The cascade code of the internal encoding n_1 and external encoding n_2 can be denoted as the concatenated code $n_1 \times n_2$. The internal encoding is a qubit code $C_1 = (E^1, D^1)$ and the external encoding is a qubit code $C_2 = (E^2, D^2)$. Map $\mathrm{E}_1 \otimes \ldots\ldots \otimes \mathrm{E}_1 = (\mathrm{E}_1)^{\otimes n_2}$ to $\mathrm{E}^2[\rho_0]$ and the decoding mapping of concatenated code is expressed as

$$D' = D^2 \circ D^1 \bigotimes n_2. \tag{3.97}$$

Then the quantum concatenated code of $n_1 \times n_2$ is written as

$$C^2(C^1) = (E', D'). \tag{3.98}$$

In quantum concatenated code, the internal encoding is arbitrary code (stable subcode or graph state code) to avoid the occurrence of error information, while the external encoding is quantum concatenated code to prevent the occurrence of quantum error. The graph state is the pure form of n quantum bit and the stable subcode is one of the graph state codes. The application of graph state is widely used in quantum computation and communication.

The definition of the graph is

$$G = (V, E) \tag{3.99}$$

where V is the set of vertices of $\{1, \ldots\ldots, N\}$ and E is the set of edges. If a and b are two endpoints of the same edge, then a and b are adjacent. According to the adjacency relation of vertices in the graph G, we define an adjacency matrix, namely, $N \times N$ matrix, and its matrix elements are described as follows:

$$F = 1\{a, b\} \in E \text{ and } F = 0\{a, b\} \in E. \tag{3.100}$$

The set of points adjacent to the vertex a is

$$N_a := \{b \in V \mid \{a, b\} \in E\}. \tag{3.101}$$

Each vertex in the graph corresponds to a qubit and the graph states correspond to the pure states of qubits in Hilbert space $H = (C^2)^{\otimes N}$.

The pure state of the graph state r is defined as

$$|G\rangle = \prod_{F_{ab}=1} U_{ab} \mid +\rangle_x^v = \frac{1}{\sqrt{2}} \sum_{\mu=0}^{1} (-1)^{\frac{1}{2}\mu^F \mu^T} \mid \mu\rangle \tag{3.102}$$

where $|\mu\rangle$ is the joint eigenstate and U_{ab} is the control phase gate acting on quantum bits a and b. The expression is

$$U_{ab} = \begin{bmatrix} 1 & 0 & 0 & 0 \\ 0 & 1 & 0 & 0 \\ 0 & 0 & 1 & 0 \\ 0 & 0 & 0 & -1 \end{bmatrix}. \tag{3.103}$$

The graph state basis, characterized by its orthogonal normalization, can facilitate the block diagonalization. Then can reduce the complexity of calculation. The graph state basis is defined as $|G_{k_1 k_2,\ldots,k_n}\rangle = \prod_{a \in V} Z_a^{k_a} \mid G\rangle, K_a = 0, 1$. Based on the characteristics of graph state basis, we can get

$$k_a |G_{k_1,k_2,\ldots\ldots,k_n}\rangle = (-1)^{k_a} |G_{k_1,k_2,\ldots\ldots\ldots k_n}\rangle. \tag{3.104}$$

Therefore, we can analyze the concatenated code by graph state. If the outer graph state is $|W\rangle$, then $W = (V; F)$. And if the inter-graph state is $|w\rangle$, then $w = (v;])$. Therefore, the quantum concatenated code can be expressed as $w \otimes W$ and the concatenated graph state can be expressed as

$$\frac{1}{\sqrt{2^{Vv}}} \sum_{\mu_1 \mu_2 \ldots \mu_n} (-1)^{\sum_{i<j}^{F_{ij}\mu_j^T}} |h_1\rangle |h_2\rangle \ldots\ldots |h_n\rangle. \tag{3.105}$$

When $\Gamma = 0$, the concatenated graph state is expressed as

$$\frac{1}{\sqrt{2^n}} \sum_w \mid W\rangle = \mid g\rangle^{\otimes n}. \tag{3.106}$$

In graph state coding, the graph diagonal state of the input state ρ^{in} is $\rho^{in} = \sum_k \pi_k | G_k\rangle\langle G_k|$. In the representation of the Kraus sum operator, the effect of a single Pauli channel on a single qubit is

$$N(\rho) = f_\rho + P_X X_1 \rho X_2 + P_y Y_1 \rho Y_1 + P_z Z_1 \rho Z_1 \tag{3.107}$$

where f is the fidelity and X_1, X_2, and X_3 are the Pauli operators.

The output state of the channel is

$$\sigma^c = N^{\bigotimes n}\left(\rho^{in}\right) = \sum_a \eta_a E_a \rho^c E_a^\dagger. \tag{3.108}$$

The matrix element of the output state of the channel under the graph state basis is

$$\sigma_x^c = \sum_k \pi_k \sum_a \eta_a < G|E_a|G_k\rangle \left\langle G_k\left|E_a^\dagger\right|G\right\rangle = \delta_x \sum_k \pi_k \sum_{a|E_a \in Z^k \bigotimes l_k} \eta_a. \tag{3.109}$$

The joint output state of the channel is

$$\sigma^{AC} = \sum_{ij} \sqrt{\pi_i \pi_j} \sum_a \eta_a \left(|G_i\rangle_A \left\langle G_j\right|_A\right) \bigotimes \left(E_a |G_i\rangle \left\langle G_j\right| E_a^\dagger\right). \tag{3.110}$$

Under the graph state basis, it can be expressed as

$$\sigma^{AC\prime} = \left\langle G_s\left|A\left\langle G_m\right|\sigma^{AC}\right|G_t\right\rangle A\left|G_1\right\rangle = \sum_{ij} \sqrt{\pi_i \pi_j} \sum_a \eta_a < G_m|E_a|G_i\rangle \left\langle G_j\left|E_a^\dagger\right|G\right\rangle \delta_i \delta_j. \tag{3.111}$$

According to the orthogonality of the graph state basis, we can get

$$\sigma_i^{AC} = \sqrt{\pi_i \pi_j} \sum_{a|Ea \in Z^k k} (-1) P_a \eta_a. \tag{3.112}$$

3.3.2 Channel Capacity Coding of Graph State Basis

We use quantum graph state concatenated code, information security theory, and quantum random coding theory to encode tree graph and forest graph with multiple degrees of freedom in immune noise channel by graph state basis method.

3.3.2.1 Channel Coding Based on Tree Graph

For the concatenated code of $n \times n$, when $n_2 = 1$ and the logical basis group $|\overline{0}\rangle = |G\rangle$, $|\overline{1}\rangle = Z_n|G\rangle$ is chosen, then the total input quantum state is

$$\rho^c = \frac{1}{2}\left(|\overline{0}\rangle\langle\overline{0}| + |\overline{1}\rangle\langle\overline{1}|\right) = \frac{1}{2}\left(|G\rangle\langle G| + |G_{00\ldots1}\rangle\langle 00\ldots1|\right). \tag{3.113}$$

If $|\overline{0}\rangle$ and $|\overline{1}\rangle$ are bases, then the joint output state σ^{AC} can be partitioned and each matrix after partitioning is

$$\frac{1}{2}\begin{bmatrix} \sum_a \eta_a + \eta_0 & \sum_a \eta_a - \eta_0 \\ \sum_a \eta_a - \eta_0 & \sum_a \eta_a + \eta_0 \end{bmatrix}. \tag{3.114}$$

When the coset leader of the error operator is $Z_1Z_2 \ldots Z_i (0 \leq i \leq n_1 - 1)$, then

$$\begin{aligned} \eta_e = a_i = \frac{1}{2}\Big[&(P_z + P_y)^i (f + P_z)(f + P_x)^{n_1 - i - 1} \\ &+ (P_z - P_y)^i (f - P_z)(f - P_x)^{n_1 - i - 1}\Big] \end{aligned}. \tag{3.115}$$

$$\begin{aligned} b_i = \frac{1}{2}\Big[&(f + P_x)^i (P_x + P_y)(P_z + P_y)^{n_1 - i - 1} \\ &+ (f - P_x)^i (P_x - P_y)(P_z - P_y)^{n_1 - i - 1}\Big] \end{aligned}. \tag{3.116}$$

When the coset head of the error operator is $Z_1Z_2 \ldots Z_i Z_n (0 \leq i \leq n_1 - 1)$ then

$$\begin{aligned} \eta_e = a_i = \frac{1}{2}\Big[&(P_z + P_y)^i (f + P_z)(f + P_x)^{n_1 - i - 1} \\ &- (P_z - P_y)^i (f - P_z)(f - P_x)^{n_1 - i - 1}\Big] \end{aligned}. \tag{3.117}$$

$$\begin{aligned} b_i = \frac{1}{2}\Big[&(f + P_x)^i (P_x + P_y)(P_z + P_y)^{n_1 - i - 1} \\ &- (f - P_x)^i (P_x - P_y)(P_z - P_y)^{n_1 - i - 1}\Big] \end{aligned}. \tag{3.118}$$

The eigenvalue of the output state σ^c is

$$\sigma^c = \frac{1}{2}[\eta_e(k) + \eta_0(k)] + \frac{1}{2}\Big[\eta_e\big(k \bigoplus 0.....01\big) + \eta_0\big(k \bigoplus 0.....01\big)\Big]. \tag{3.119}$$

For the tree graph, the coherent information of the channel is

$$\begin{aligned} I\big(\rho^c, N^{\bigotimes N_1}\big) &= S(\sigma^c) - S(\sigma^{AC}) \\ &= \sum_{i=0}^{n_1 - 1} C_{n_1 - 1}^i (a_i \log_2 a_i + b_i \log_2 b_i + c_i \log_2 c_i \\ &+ d_i \log_2 d_i - a_i + b_i + c_i + d_i) \log_2 \frac{1}{2}(a_i + b_i + c_i + d_i). \end{aligned} \tag{3.120}$$

3.3.2.2 Channel Coding Based on Forest Graph

For the forest graph of $n_1 \times n_2$, the subgroup k is divided into two subsets $k(0)$ and $k(e)$. For the coset head $E = Z^k K$, Z^k consists of n_2 parts, and each part corresponds to the coset head of the tree graph. For a particular element of the group K, assuming that in l_i and s there are l_i^0 and s_i^n parts which anticommuting to it. Then the contribution value of this element to the eigenvalue of joint output states is

$$\prod_{i=0}^{n_1-1} a_i^{l_i - l_i^0} b_i^{l_i^0} c_i^{s_i - s_i^0} d_i s_i^0 \tag{3.121}$$

Summing over all the elements in the group K, we get

$$\eta^e(k) \pm \eta^0(k) = \prod_{i=0}^{n_1-1} (a_i \pm b_i)^{l_i} (c_i \pm d_i)^{s_i} \tag{3.122}$$

Therefore, we know that

$$\eta^e(k) = \frac{1}{2}\left(\prod_{i=0}^{n_1-1} (a_i + b_i)^{l_i} (c_i + d_i)^{s_i} \pm \prod_{i=0}^{n_1-1} (a_i - b_i)^{l_i} (c_i - d_i)^{s_i}\right) \tag{3.123}$$

Analyzing from the forest graph, we can get

$$\eta^e\left(k \bigoplus 1'\right) + \eta^0\left(k \bigoplus 1'\right) = \prod_{i=0}^{n_1-1} (a_i + b_i)^{s_i} (c_i + d_i)^{l_i} \tag{3.124}$$

The eigenvalue of the output state is

$$\begin{aligned} \eta' &= \frac{1}{2}\left[\eta^e(k) + \eta^0(k) + \eta^e\left(k \bigoplus 1'\right) + \eta^0\left(k \bigoplus 1'\right)\right. \\ &= \frac{1}{2}\left[\prod_{i=0}^{n_1-1} (a_i + b_i)^{l_i} (c_i + d_i)^{s_i} + \prod_{i=0}^{n_1-1} (a_i + b_i)^{s_i} (c_i + d_i)^{l_i}\right] \end{aligned} \tag{3.125}$$

The degeneracy is known as

$$\Delta = n_2 \prod_{i=0}^{n_2-1} \frac{1}{l_i s_i} c_{n_1-1}^{i \quad l_i + s_i} \tag{3.126}$$

For the forest graph of $n_1 \times n_2$, the average coherent information of each channel is

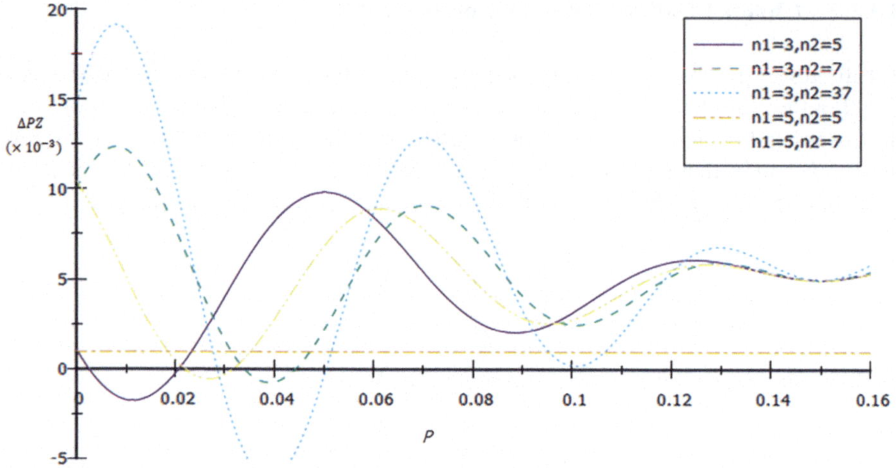

Fig. 3.29 The noise tolerance when $\theta = 0$

$$I^{CN} = \frac{1}{n_1 n_2} \sum \{ \Delta [-\eta' \log_2(\eta') + \eta_e \log_2(\eta_e) + \eta_0 \log_2(\eta_0)] \} \tag{3.127}$$

3.3.2.3 Analysis of Noise Tolerance

For different concatenated code can calculate the noise tolerance of the channel conveniently, when P_r, P_{ij} are known and $I > 0$, we can get the maximum value of $P(z)$. In $\tan\theta = \frac{P_y}{P_x}$, take the noise tolerance of θ at different angles.

The noise tolerance, when $\theta = 0$, is shown in Fig. 3.29:

The noise tolerance, when $\theta = \pi/12$, is shown in Fig. 3.30:

The noise tolerance, when $\theta = \pi/4$, is shown in Fig. 3.31:

We know from our analysis that concatenated code is a very effective quantum code, which can improve the quantum tolerance of communication channel. When P_x, P_y, and P_z of the channel are known, the most efficient coding method is selected by calculating the noise tolerance of different concatenated code. Compared with the previous coding methods, the coding of graph state basis we designed made full use of the orthogonal normalization property of graph state basis to diagonalize large matrices or blocks. It is proved that the computational speed is improved, and the depolarized channel, namely, the channel of $P_x = P_y = P_z$, can be improved by 3–4 orders of magnitude.

3.4 Summary of This Chapter

This chapter constructs the framework of entangled quantum teleportation channel with high fidelity. The framework characterizes the quantum entanglement evolution model in local area under the noise for the "entanglement death" problem and

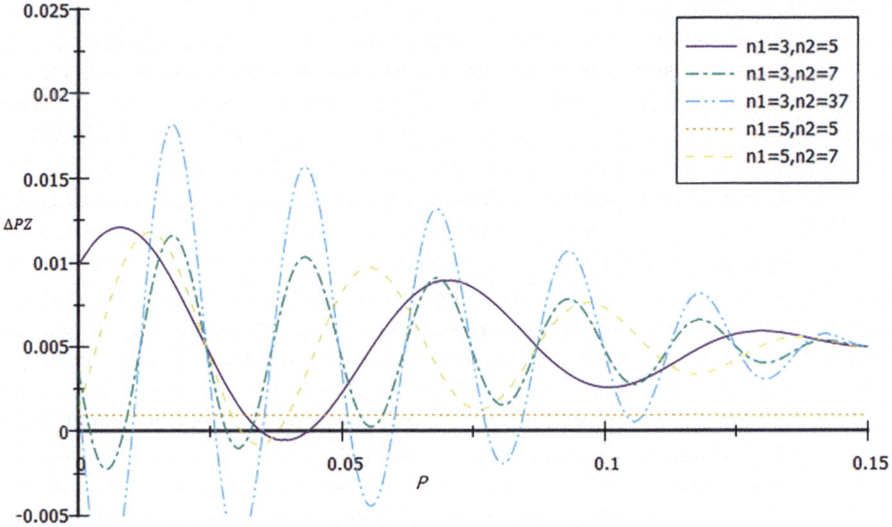

Fig. 3.30 The noise tolerance when $\theta = \pi/12$

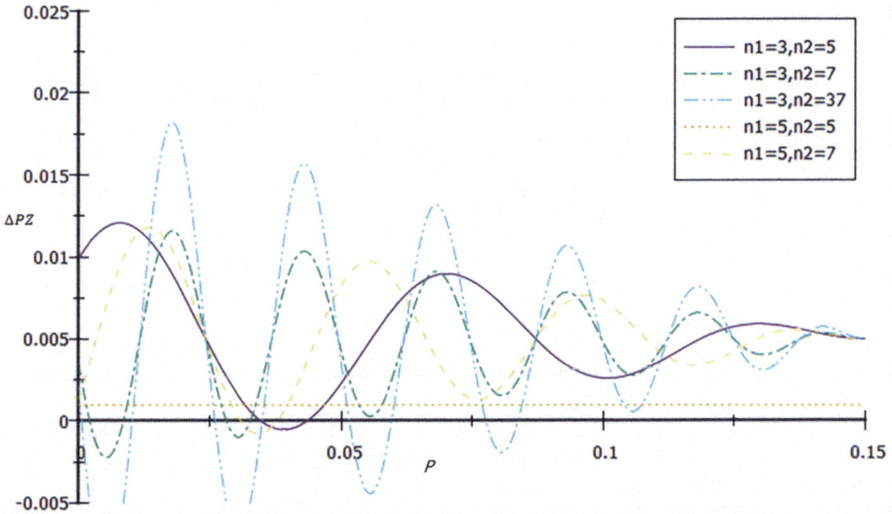

Fig. 3.31 The noise tolerance when $\theta = \pi/4$

decoherence phenomenon in the process of quantum teleportation. Based on the model, the immune noise model based on density matrix and collective noise immune model based on DFS are constructed. The channel capacity is coded based on graph state basis under noise. Firstly, the construction of quantum entanglement evolution model including the characterization of the "entanglement death" problem, which analyzes the parameters related to the entanglement degree when the entanglement death occurs through the entanglement evolution characteristics of two independent atomic systems in space and the entanglement evolution characteristics of two three-level atomic systems over time. Quantum decoherence characterization mainly analyzes quantum decoherence between atom and cavity field in T-C model and quantum decoherence between two cavity fields and two two-level atoms in J-C model. The evolution of quantum entanglement in locally common mode is characterized by the sudden death and resurrection of entanglement when the quantum channel is affected by locally common quantum noise environment. Secondly, the immune noise model based on density matrix and the immune collection noise model based on DFS are constructed. Finally, the tree graph and forest graph are coded with multi-degrees of freedom by graph state basis method, the coherence information and channel capacity are analyzed, the channel capacity under different noise channels is obtained, and the approximate value of noisy quantum capacity, calculation speed, and noise tolerance of quantum information of channel are calculated effectively. This chapter provides a unified security channel for the content of the following sections.

References

1. C. Elliott, A. Colvin, D. Pearson, et al., Current status of the DARPA quantum network, in *Proceedings of SPIE, Orlando, FL, USA*, vol. 5815, (2005), pp. 138–149
2. A. Poppe, M. Peev, O. Maurhart, Outline of the SECOQC quantum-key-distribution network in Vienna. Int. J. Quantum. Inf. **6**(2), 209–218 (2008)
3. 许方星, 陈巍, 王双, 等 (F. Xu, W. Chen, S. Wang et al.), 多层级量子密码城域网 (Multi-tiered Quantum Cryptography Metropolitan Area Network). 科学通报 (Chin. Sci. Bull.) **54**(16), 2277–2283 (2009)
4. 廖庆洪 (Q. Liao), 孤立原子和双模腔内原子之间的纠缠突然死亡研究 (Study on Sudden Death of Entanglement Between Isolated Atoms and Atoms in Bimodal Cavity). 南昌:南昌大学(Nanchang University, Nanchang, 2012), pp. 757–761
5. 魏巧 (Q. Wei), 两个V型三能级原子系统的纠缠突然死亡与复苏 (Sudden Death and Revival of Entanglement in Two V-type Three-level Atomic Systems). 武汉:武汉大学 (Wuhan University, Wuhan. 2010), pp. 4453–4459
6. 张玥 (Y. Zhang), 2个原子与2个耗散腔场量子体系中的量子退相干 (Quantum Decoherence in Quantum Systems of Two Atoms and Two Dissipative Cavity Fields). 长春:东北师范大学 (Northeast Normal University, Changchun, 2012), pp. 65–68
7. 王海红 (H. Wang), 原子相干效应中的量子噪声特性研究 (Study on Quantum Noise Characteristics in Atomic Coherence Effects) 太原:山西大学 (Shanxi University, Taiyuan, 2006), pp. 12–15

Chapter 4
Quantum Information Splitting of Bell States and Arbitrary States in Different Channels

Based on the immune noise high-capacity fault-tolerant quantum teleportation channel discussed in Chap. 3. In this chapter, according to the different characteristics of qubits, quantum entanglement is used to accomplish the same task with fewer quantum information resources or to accomplish more tasks with the same quantum information resources. Different quantum channels (e.g., entangled state, cluster state, W state, etc.) are selected for information splitting of single qubit, two qubits, and three qubits, and quantum information splitting schemes are designed to achieve Bell state and arbitrary state in different channels. At the same time, it provides the basis for designing the multi-freedom quantum teleportation protocol with immune noise and the quantum cryptography application of continuous variables.

The structure of this chapter is as follows: Firstly, Sect. 4.1 proposes the quantum information splitting based on the Bell state. Secondly, in Sect. 4.2 the five-qubit entangled states are used as quantum channels to split arbitrary single-qubit state and arbitrary two-qubit states. Thirdly, in Sect. 4.3 the four-qubit cluster state and two-qubit Bell state are used as the quantum channel to split any two-qubit state. Fourthly, in Sect. 4.3.4 the four-qubit cluster state and GHZ state are used as the quantum channel to split arbitrary three-qubit states. Finally, Sect. 4.4 summarizes the content of this chapter. Figure 4.1 shows the organizational structure of quantum information splitting.

4.1 Based on the Bell State

Suppose that the sender Alice and receivers Bob and Charlie share the four-qubit entangled state as the quantum channel, and the channel is described as follows:

D. Li, *Quantum Teleportation Under Noise*,
https://doi.org/10.1007/978-3-031-91250-4_4

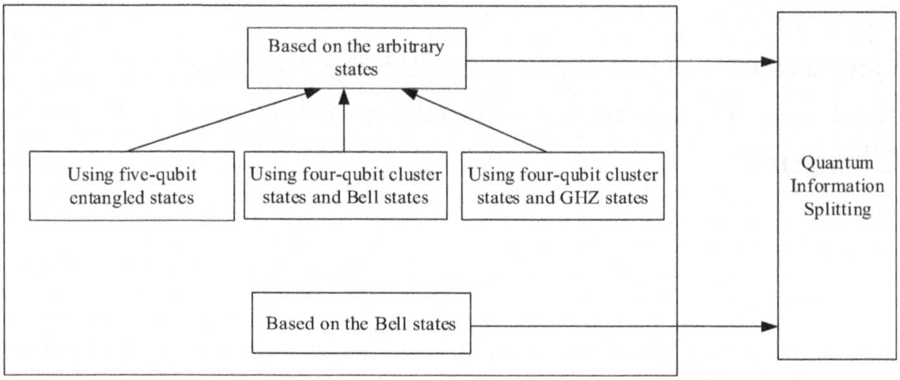

Fig. 4.1 The organizational structure of quantum information splitting

$$
\begin{aligned}
|\Phi\rangle_{1234} &= \frac{1}{2}[|00\rangle(|00\rangle+|11\rangle)+|11\rangle(|01\rangle+|10\rangle)]_{1234} \\
&= \frac{1}{2}(|0000\rangle+|0011\rangle+|1101\rangle+|1110\rangle)_{1234.}
\end{aligned} \tag{4.1}
$$

Alice's Bell state is as follows:

$$
\begin{aligned}
|\Phi^{+}\rangle_{x_1x_2} &= \frac{1}{\sqrt{2}}(|00\rangle+|11\rangle) \quad |\Phi^{-}\rangle_{x_1x_2} = \frac{1}{\sqrt{2}}(|00\rangle-|11\rangle), \\
|\Psi^{+}\rangle_{x_1x_2} &= \frac{1}{\sqrt{2}}(|01\rangle+|10\rangle) \quad |\Psi^{-}\rangle_{x_1x_2} = \frac{1}{\sqrt{2}}(|01\rangle-|10\rangle).
\end{aligned} \tag{4.2}
$$

4.1.1 Quantum Information Splitting Process

Suppose Alice has qubit 1 and two-qubit Bell states, Bob has qubits 2 and 3, and Charlie has qubit 4; then the quantum system is

$$
\begin{aligned}
|\Pi\rangle_{x_1x_21234} &= |\Phi^{\pm}\rangle_{x_1x_2} \otimes |\Phi\rangle_{1234} \\
&= \frac{1}{2\sqrt{2}}(|000000\rangle + |000011\rangle + |001101\rangle + |001110\rangle \\
&\pm |110000\rangle \pm |110011\rangle \pm |111101\rangle \pm |111110\rangle)_{x_1x_21234}
\end{aligned}
$$

or

$$
\begin{aligned}
|\Pi\rangle_{x_1x_21234} &= |\Psi^{\pm}\rangle_{x_1x_2} \otimes |\Phi\rangle_{1234} \\
&= \frac{1}{2\sqrt{2}}(|010000\rangle + |010011\rangle + |011101\rangle + |011110\rangle \pm |100000\rangle \\
&\pm |100011\rangle \pm |101101\rangle \pm |101110\rangle)_{x_1x_21234}.
\end{aligned}
$$

The sender Alice has the initial states $|\Phi^{\pm}\rangle_{x_1x_2}$ and $|\Psi^{\pm}\rangle_{x_1x_2}$. To realize the quantum information splitting, Alice will choose one of Bob and Charlie to reconstruct the original state. Therefore, Alice will first perform the measurement of the GHZ state of her qubit 1, x_1, x_2, and the measurement basis is

$$
\begin{aligned}
|\mathrm{GHZ}\rangle^{\pm} &= \frac{1}{\sqrt{2}}(|000\rangle \pm |111\rangle), \quad |H\rangle^{\pm} = \frac{1}{\sqrt{2}}(|011\rangle \pm |100\rangle), \\
|G\rangle^{\pm} &= \frac{1}{\sqrt{2}}(|010\rangle \pm |101\rangle), \quad |Z\rangle^{\pm} = \frac{1}{\sqrt{2}}(|001\rangle \pm |110\rangle).
\end{aligned}
\tag{4.3}
$$

When Alice completes her measurement, Bob and Charlie's qubits will collapse into one of the following equations:

$$
\begin{aligned}
|\Psi^{1,2}\rangle_{234} &= \frac{1}{4}(|000\rangle + |011\rangle \pm |101\rangle \pm |110\rangle)_{234}, \\
|\Psi^{3,4}\rangle_{234} &= \frac{1}{4}(|101\rangle + |110\rangle \pm |000\rangle \pm |011\rangle)_{234}, \\
|\Psi^{5,6}\rangle_{234} &= \frac{1}{4}(|000\rangle + |011\rangle \mp |101\rangle \mp |110\rangle)_{234}, \\
|\Psi^{7,8}\rangle_{234} &= \frac{1}{4}(|101\rangle + |110\rangle \mp |000\rangle \mp |011\rangle)_{234}.
\end{aligned}
\tag{4.4}
$$

According to the principle of quantum entanglement, the quantum information sent by Alice has been transmitted to the quantum states of the recipients Bob and Charlie. Then, Alice sends the measurement results to the recipient Bob (Charlie) via the classical channel, if Bob (Charlie) only makes local measurements of the qubits he owns, he can't get the initial state that Alice sends, and according to quantum no-cloning theorem, he must work with controller Charlie (Bob) to get the message sent by Alice.

Assuming that Alice's measurement result is $|\Psi^2\rangle_{234}$, according to Formula (4.6), the collapse state of Bob and Charlie is

$$
\begin{aligned}
|\Psi^2\rangle_{234} &= \frac{1}{4}(|000\rangle + |011\rangle - |101\rangle - |110\rangle) \\
&= \frac{1}{\sqrt{2}}[|0\rangle_2(|00\rangle + |11\rangle)_{34}] - |1\rangle_2(|01\rangle + |10\rangle)_{34}.
\end{aligned}
\tag{4.5}
$$

For the collapse state in Formula (4.4), Bob needs to perform the unitary operation to recover the initial state, which is expressed as

$$I = I = U_0 = |0\rangle\langle 0| + |1\rangle\langle 1|, \qquad Z = \sigma_z = U_1 = |0\rangle\langle 0| - |1\rangle\langle 1|,$$
$$X = \sigma_x = U_2 = |0\rangle\langle 1| + |1\rangle\langle 0|, \quad Y = -i\sigma_u = U_3 = |0\rangle\langle 1| - |1\rangle\langle 0|. \tag{4.6}$$

If Alice asks Bob to restore to the initial state and Charlie is willing to cooperate with Bob, Charlie needs to perform the single-qubit measurement and inform Bob of the results when the measurement is complete.

If Charlie agrees that Bob reconstructs the initial state, he needs to perform the single-qubit measurement on qubit 4 with $\{|0\rangle, |1\rangle\}$ as the base. After Charlie's measurement is completed, the qubit owned by Bob will collapse, and the collapsed state will be one of the following:

$$\begin{aligned} |\xi^{1,2}\rangle &= (|00\rangle \pm |11\rangle)_{23}\ |\xi^{3,4}\rangle = (|01\rangle \pm |10\rangle)_{23}, \\ |\xi^{5,6}\rangle &= (|11\rangle \pm |00\rangle)_{23}\ |\xi^{7,8}\rangle = (|10\rangle \pm |01\rangle)_{23}, \\ |\xi^{9,10}\rangle &= (|00\rangle \mp |11\rangle)_{23}\ |\xi^{11,12}\rangle = (|01\rangle \mp |10\rangle)_{23}, \\ |\xi^{13,14}\rangle &= (|11\rangle \mp |00\rangle)_{23}\ |\xi^{15,16}\rangle = (|10\rangle \mp |01\rangle)_{23}. \end{aligned} \tag{4.7}$$

If Charlie's measurement result is $|0\rangle_4$, then Bob's collapsed state is $(|00\rangle + |11\rangle)_{23}$. This is the initial information sent by Alice, so Bob doesn't need to do anything to his qubits to obtain the original state. If Charlie's measurement result is $|1\rangle_4$, then Bob's collapsed state is $(|01\rangle + |10\rangle)_{23}$. Then Bob only needs to perform operation U_2 on qubits 2 and 3 to get the initial information sent by Alice.

After Bob gets the measurement results of Alice and Charlie, he can perform the corresponding unitary operation to obtain the initial states $|\Phi^{\pm}\rangle_{x_1x_2}$ and $|\Psi^{\pm}\rangle_{x_1x_2}$.

The quantum information splitting process of the two-qubit Bell state is shown in Fig. 4.2.

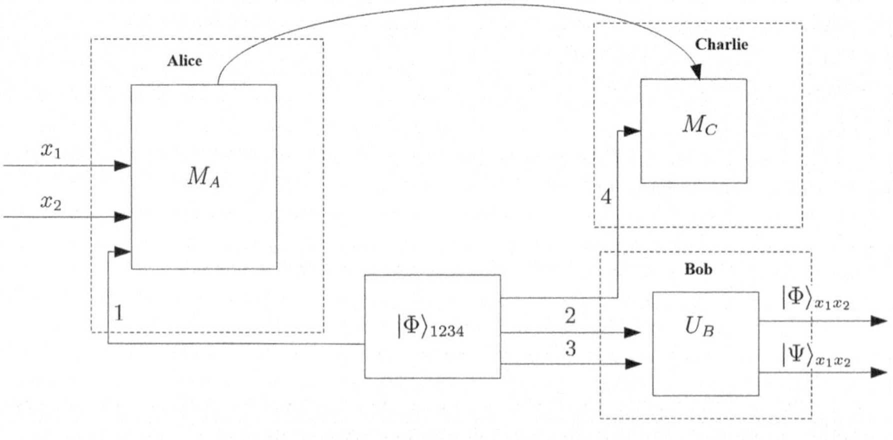

Fig. 4.2 Quantum information splitting process of two-qubit Bell state

To complete the quantum information splitting, Alice performs the GHZ state measurement of qubit 1, $x1$, $x2$, and after the measurement is completed, the qubits of Bob and Charlie will collapse, and the states obtained by Bob, the single-qubit measurement performed by Charlie, and the unitary operation performed by Bob are shown in Table 4.1.

4.1.2 Security Analysis

This book analyzes security from internal and external aspects.

Suppose there is an eavesdropper Eve in the process of quantum information splitting, who tries to entangle an auxiliary qubit in the quantum channel to obtain the initial state information sent by Alice, and Alice, Bob, and Charlie didn't find it during the eavesdropping, so when Alice made GHZ state measurements of the qubits she had and the ones she was going to transmit, the quantum state owned by Alice, Bob, Charlie, and the eavesdropper Eve would collapse. Charlie then makes the single-qubit measurement on his qubit, so that Bob and Eve's system collapses into a direct product state, where Eve will not have any information about the original state.

Further, suppose eavesdropper Eve wants to entangle auxiliary qubit E to Charlie's qubit 2, which is represented as $\frac{1}{\sqrt{2}}(|0\rangle+|1\rangle)_E$.

If the measurement result obtained by Alice is $|\Psi^i\rangle_{234E}(i=1,2,\ldots\ldots,8)$, the joint state obtained by Bob, Charlie, and the eavesdropper Eve is

$$\begin{aligned}
|\Omega^{1,2}\rangle_{234E} &= \frac{1}{4}(|0000\rangle + |0110\rangle \pm |1010\rangle \pm |1100\rangle)_{234E},\\
|\Omega^{3,4}\rangle_{234E} &= \frac{1}{4}(|0001\rangle + |0111\rangle \pm |1011\rangle \pm |1101\rangle)_{234E},\\
|\Omega^{5,6}\rangle_{234E} &= \frac{1}{4}(|1010\rangle + |1100\rangle \pm |0000\rangle \pm |0110\rangle)_{234E},\\
|\Omega^{7,8}\rangle_{234E} &= \frac{1}{4}(|1011\rangle + |1101\rangle \pm |0001\rangle \pm |0111\rangle)_{234E},\\
|\Omega^{9,10}\rangle_{234E} &= \frac{1}{4}(|0000\rangle + |0110\rangle \mp |1010\rangle \mp |1100\rangle)_{234E},\\
|\Omega^{11,12}\rangle_{234E} &= \frac{1}{4}(|0001\rangle + |0111\rangle \mp |1011\rangle \mp |1101\rangle)_{234E},\\
|\Omega^{13,14}\rangle_{234E} &= \frac{1}{4}(|1010\rangle + |1100\rangle \mp |0000\rangle \mp |0110\rangle)_{234E},\\
|\Omega^{15,16}\rangle_{234E} &= \frac{1}{4}(|1011\rangle + |1101\rangle \mp |0001\rangle \mp |0111\rangle)_{234E}.
\end{aligned} \tag{4.8}$$

When Charlie performs the single-qubit measurement, the quantum state after the Bob-Eve collapse is

Table 4.1 Operations performed by Alice, Bob, and Charlie

Alice's measurement	The state obtained by Bob and Charlie	Charlie's measurement	States obtained by Bob	Unitary operators (U_i)
$\|GHZ\rangle^+$	$\|\Psi^1\rangle_{234} = \frac{1}{4}(\|000\rangle + \|011\rangle + \|101\rangle + \|110\rangle)_{234}$	$\|0\rangle$	$\|\xi^1\rangle = (\|00\rangle + \|11\rangle)_{23}$	U_0
		$\|1\rangle$	$\|\xi^2\rangle = (\|00\rangle - \|11\rangle)_{23}$	U_1
$\|GHZ\rangle^-$	$\|\Psi^2\rangle_{234} = \frac{1}{4}(\|000\rangle + \|011\rangle - \|101\rangle + \|110\rangle)_{234}$	$\|0\rangle$	$\|\xi^3\rangle = (\|01\rangle + \|10\rangle)_{23}$	U_2
		$\|1\rangle$	$\|\xi^4\rangle = (\|01\rangle - \|10\rangle)_{23}$	U_3
$\|G\rangle^+$	$\|\Psi^3\rangle_{234} = \frac{1}{4}(\|101\rangle + \|110\rangle + \|000\rangle + \|011\rangle)_{234}$	$\|0\rangle$	$\|\xi^5\rangle = (\|11\rangle + \|00\rangle)_{23}$	U_0
		$\|1\rangle$	$\|\xi^6\rangle = (\|11\rangle - \|00\rangle)_{23}$	$-U_1$
$\|G\rangle^-$	$\|\Psi^4\rangle_{234} = \frac{1}{4}(\|101\rangle + \|110\rangle - \|000\rangle - \|011\rangle)_{234}$	$\|0\rangle$	$\|\xi^7\rangle = (\|10\rangle + \|01\rangle)_{23}$	U_2
		$\|1\rangle$	$\|\xi^8\rangle = (\|10\rangle - \|01\rangle)_{23}$	$-U_3$
$\|H\rangle^+$	$\|\Psi^5\rangle_{234} = \frac{1}{4}(\|000\rangle + \|011\rangle - \|101\rangle - \|110\rangle)_{234}$	$\|0\rangle$	$\|\xi^9\rangle = (\|00\rangle - \|11\rangle)_{23}$	U_1
		$\|1\rangle$	$\|\xi^{10}\rangle = (\|00\rangle + \|11\rangle)_{23}$	U_0
$\|H\rangle^-$	$\|\Psi^6\rangle_{234} = \frac{1}{4}(\|000\rangle + \|011\rangle + \|101\rangle + \|110\rangle)_{234}$	$\|0\rangle$	$\|\xi^{11}\rangle = (\|01\rangle - \|10\rangle)_{23}$	U_3
		$\|1\rangle$	$\|\xi^{12}\rangle = (\|01\rangle + \|10\rangle)_{23}$	U_2
$\|Z\rangle^+$	$\|\Psi^7\rangle_{234} = \frac{1}{4}(\|101\rangle + \|110\rangle - \|000\rangle - \|011\rangle)_{234}$	$\|0\rangle$	$\|\xi^{13}\rangle = (\|11\rangle - \|00\rangle)_{23}$	$-U_1$
		$\|1\rangle$	$\|\xi^{14}\rangle = (\|11\rangle + \|00\rangle)_{23}$	U_0
$\|Z\rangle^-$	$\|\Psi^8\rangle_{234} = \frac{1}{4}(\|101\rangle + \|110\rangle + \|000\rangle + \|011\rangle)_{234}$	$\|0\rangle$	$\|\xi^{15}\rangle = (\|10\rangle - \|01\rangle)_{23}$	$-U_3$
		$\|1\rangle$	$\|\xi^{16}\rangle = (\|10\rangle + \|01\rangle)_{23}$	U_2

$$|\Xi^{1,2}\rangle=\frac{1}{4}(|000\rangle+|010\rangle\pm|100\rangle\pm|110\rangle)_{23E}|\Xi^{3,4}\rangle=\frac{1}{4}(|001\rangle+|011\rangle\pm|101\rangle\pm|111\rangle)_{23E,}$$

$$|\Xi^{5,6}\rangle=\frac{1}{4}(|100\rangle+|110\rangle\pm|000\rangle\pm|010\rangle)_{23E}|\Xi^{7,8}\rangle=\frac{1}{4}(|101\rangle+|111\rangle\pm|001\rangle\pm|011\rangle)_{23E,}$$

$$|\Xi^{9,10}\rangle=\frac{1}{4}(|000\rangle+|010\rangle\mp|100\rangle\mp|110\rangle)_{23E}|\Xi^{11,12}\rangle=\frac{1}{4}(|001\rangle+|011\rangle\mp|101\rangle\mp|111\rangle)_{23E,}$$

$$|\Xi^{13,14}\rangle=\frac{1}{4}(|100\rangle+|110\rangle\mp|000\rangle\mp|010\rangle)_{23E}|\Xi^{15,16}\rangle=\frac{1}{4}(|101\rangle+|111\rangle\mp|001\rangle\mp|011\rangle)_{23E.}$$

Therefore, when Charlie's measurement result is $|0\rangle$, Bob-Eve will collapse to $|\beth^1\rangle = (|00\rangle + |01\rangle + |10\rangle + |11\rangle)_{BE}$. It follows that Eve does not obtain any information about the original state, which means that the information splitting scheme we construct is safe and reliable.

Suppose one of the three communicating parties is dishonest, and suppose Charlie is dishonest or he's working with the eavesdropper Eve. When Alice sends the initial state information, Charlie will intercept the quantum information sent by Alice to Bob, and if Alice lets Charlie reconstruct the original state, Charlie's eavesdropping behavior will be realized without being detected, but if Alice lets Bob reconstruct the initial state, since Charlie does not know Alice's measurement results, therefore, he may send the wrong quantum state information to Bob. From the above analysis, it is clear that Eve cannot obtain any information about the original state, which shows eavesdropping behavior will also be discovered.

4.2 Using the Five-Qubit Entangled State to Split Arbitrary Single-Qubit and Two-Qubit State

The five-particle entanglement states are as follows:

$$\begin{aligned}|\Theta\rangle 12345 &= \frac{1}{2\sqrt{2}}(|00101\rangle - |00110\rangle + |01000\rangle - |01011\rangle + |10001\rangle + |10010\rangle \\ &\quad + |11100\rangle + |11111\rangle)_{12345} \\ &= \frac{1}{2}(|001\rangle|\Phi^-\rangle + |010\rangle|\Psi^-\rangle + |100\rangle|\Phi^+\rangle + |111\rangle|\Psi^+\rangle)_{12345.}\end{aligned} \tag{4.9}$$

The five-qubit entangled states have been widely used in quantum one-way communication and quantum error correction, and one-way quantum teleportation has been experimentally verified and is widely used in quantum teleportation and quantum error correction codes. The following shows the process of quantum information splitting using five-qubit entangled states for arbitrary single-qubit and two-qubit state.

4.2.1 Split the Arbitrary Single-Qubit

If Alice has qubits 1, 3, and 4, Charlie has qubit 2, and Bob has qubit 5, Alice splits the arbitrary single-qubit through the quantum channel $|\Theta\rangle_{12345}$. The arbitrary single-qubit is as follows:

$$|\Psi\rangle_A = \alpha|0\rangle + \beta|1\rangle, \tag{4.10}$$

where $|\alpha|^2 + |\beta|^2 = 1$.

The whole quantum system is

$$\begin{aligned}|\Psi\rangle_{A12345} &= |\Psi\rangle_A \otimes |\Theta\rangle_{12345} = \frac{1}{2}(\alpha|0001\rangle|\Phi^-\rangle + \alpha|0010\rangle|\Psi^-\rangle \\ &+\alpha|0100\rangle|\Phi^+\rangle + \alpha|0111\rangle|\Psi^+\rangle + \beta|1001\rangle|\Phi^-\rangle + \beta|1010\rangle|\Psi^-\rangle \\ &+\beta|1100\rangle|\Phi^+\rangle + \beta|1111\rangle|\Psi^+\rangle)_{A12345}.\end{aligned} \tag{4.11}$$

Alice performs the von Neumann measurement on her qubits and gets 16 possible collapse states $|\phi\rangle^i_{A134} (i = 1, 2, 3, \ldots\ldots, 16)$. Then Alice informs Bob and Charlie of the collapsed result through the classical channel. Neither of them can reconstruct the information of the original state, and only the two parties cooperate with each other to recover the information of the unknown state. Charlie performs single-qubit measurement based on $\{|0\rangle, |1\rangle\}$ and combines the measurement results of Alice and Charlie. Bob can recover the original state information by performing appropriate unitary operations.

If Alice's measurement result is $|\phi\rangle^1_{A134}$,

$$|\phi\rangle^1_{A134} = \frac{1}{2}(|0000\rangle + |0011\rangle + |1100\rangle + |1111\rangle), \tag{4.12}$$

then the collapse state obtained by Bob and Charlie is

$$|\psi\rangle^1_{25} = (\alpha|10\rangle - \alpha|00\rangle + \beta|01\rangle + \beta|11\rangle)_{25}. \tag{4.13}$$

If Charlie performs the measurement based on $|0\rangle$, then Bob obtains $(-\alpha|0\rangle + \beta|1\rangle)_5$, so Bob can recover the information of the original state by performing the unitary operation $-\sigma_z$.

If Charlie performs the measurement based on $|1>$, then Bob obtains $(\alpha|0\rangle + \beta|1\rangle)_5$, so Bob can recover the information of the original state by performing then unitary operation I.

4.2.2 Split the Arbitrary Two Qubits

Analyze the quantum information splitting process of the arbitrary two qubits by using five-qubit entangled states, and perform the quantum control gate operation using Formula (4.9). First, with qubit 3 as the target and qubit 4 as the control qubit, the quantum state is changed after the execution of the control gate operation as

$$|\Theta'\rangle_{12345} = \frac{1}{2\sqrt{2}}(|00101\rangle + |00110\rangle + |01000\rangle - |01011\rangle + |10001\rangle + |10010\rangle + |11100\rangle - |11111\rangle)12345. \tag{4.14}$$

Then with qubit 3 as the target qubit and qubit 4 as the control qubit, performing the control CNOT operation, we write the five-qubit entangled state as

$$|\Omega\rangle_{12345} = \frac{1}{2\sqrt{2}}(|00101\rangle + |00010\rangle + |01000\rangle - |01111\rangle + |10001\rangle + |10110\rangle + |11100\rangle - |11011\rangle)_{12345}. \tag{4.15}$$

Suppose Alice wants to send an unknown two-qubit state, as follows:

$$|\varphi\rangle_{xy} = a|00\rangle + b|01\rangle + c|10\rangle + d|11\rangle, \tag{4.16}$$

where $|a|^2 + |b|^2 + |c|^2 + |d|^2 = 1$.

Alice prepares a five-qubit entangled state $|\Omega\rangle_{12345}$; qubits 1 and 2 belong to Alice, qubit 5 belongs to Charlie, and qubits 3 and 4 belong to Bob.

We write the whole quantum system as

$$\begin{aligned} |\Pi\rangle_{xy12345} &= |\varphi\rangle_{xy} \otimes |\Omega\rangle_{12345} \\ &= \frac{\sqrt{2}}{4}(a|0000101\rangle + a|0000010\rangle + a|0001000\rangle \\ &- a|0001111\rangle + a|0010001\rangle + a|0010110\rangle + a|0011100\rangle \\ &- a|0011011\rangle + b|0100101\rangle + b|0100010\rangle + b|0101000\rangle \\ &- b|0101111\rangle + b|0110001\rangle + b|0110110\rangle + b|0111100\rangle \\ &- b|0111011\rangle + c|1000101\rangle + c|1000010\rangle + c|1001000\rangle \\ &- c|1001111\rangle + c|1010001\rangle + c|1010110\rangle + c|1011100\rangle \\ &- c|1011011\rangle + d|1100101\rangle + d|1100010\rangle + d|1101000\rangle \\ &- d|1101111\rangle + d|1110001\rangle + d|1110110\rangle \\ &+ d|1111100\rangle - d|1111011\rangle)xy12345. \end{aligned} \tag{4.17}$$

First Alice performs Bell state measurements on the qubit pairs $(A, 1)(B, 2)$, and she obtains 16 possible measurements and 16 collapsed states $|\Gamma\rangle^{i}(i = 1, 2, 3, \ldots\ldots, 16)$. Alice tells the measurement result to Bob and Charlie through the classical channel, and the controller Charlie performs the single-qubit measurement on his qubit 5 based on $\{|0\rangle, |1\rangle\}$ and sends the result to Bob. Bob reconstructs the original state $|\varphi\rangle_{xy}$ by performing the appropriate unitary operation $(I, \sigma_x, \sigma_y, \sigma_z)$.
The four unitary operations are described as follows:

$$\begin{aligned} I &= |0\rangle + |1\rangle = I = \begin{bmatrix} 1 & 0 \\ 0 & 1 \end{bmatrix}, \quad X = |1\rangle + |0\rangle = \sigma_x = \begin{bmatrix} 0 & 1 \\ 1 & 0 \end{bmatrix}, \\ Y &= |1\rangle - |0\rangle = \sigma_y = \begin{bmatrix} 0 & -i \\ i & 0 \end{bmatrix}, \quad Z = |0\rangle - |1\rangle = \sigma_z = \begin{bmatrix} 1 & 0 \\ 0 & -1 \end{bmatrix}. \end{aligned} \tag{4.18}$$

Alice performs operation $|\Phi^{+}\rangle_{A1}|\Phi^{+}\rangle_{B2}$ on the qubit she has; then she gets the measurement results:

$$\begin{aligned} |\xi^{1}\rangle_{345} = \frac{1}{4\sqrt{2}}&(a|101\rangle + a|010\rangle + b|000\rangle - b|111\rangle + c|001\rangle + c|110\rangle + d|100\rangle \\ &- d|011\rangle)345. \end{aligned} \tag{4.19}$$

If Charlie's measurement result is $|0\rangle$, then Bob's state is

$$|\zeta^{1}\rangle_{34} = (a|01\rangle + b|00\rangle + c|11\rangle + d|10\rangle)_{34}. \tag{4.20}$$

Thus, Bob can restore the original state by performing unitary operation $I \otimes \sigma_x$.
If Charlie's measurement result is $|1\rangle$, then Bob's state is

$$\begin{aligned} |\zeta^{2}\rangle_{34} = (a|10\rangle - b|11\rangle \\ + c|00\rangle - d|01\rangle)_{34}. \end{aligned} \tag{4.21}$$

Thus, Bob can restore the original state by performing unitary operation $\sigma_x \otimes \sigma_z$.

4.2.3 Security Analysis

Assuming the existence of an attacker, Eve, trying to steal unknown information sent during three-party communication, and if the communicators are not aware of the presence of the eavesdropper, after Alice performs the measurement, the eavesdropper entangled his qubit in Bob and Charlie's collapsed state, which could be analyzed in two ways:

1. When splitting the arbitrary single-qubit, Alice performs the von Neumann measurement if the result obtained is

$$|\Phi^1\rangle_{A134}=\frac{1}{2}(|0000\rangle+|0011\rangle+|1100\rangle+|1111\rangle)_{A134}. \tag{4.22}$$

Then the collapse state of Eve-Bob-Charlie is

$$|\Xi^1\rangle_{25E}=(\alpha|100\rangle+\alpha|101\rangle-\alpha|000\rangle-\alpha|001\rangle+\beta|010\rangle+\beta|011\rangle+\beta|110\rangle+\beta|111\rangle)_{25E}. \tag{4.23}$$

If Charlie obtains the state $0\rangle_2$, then Eve-Bob's system will collapse to $|\Pi^1\rangle_{5E}$, which we denote as

$$|\Pi^1\rangle_{5E}=(-\alpha|00\rangle-\alpha|01\rangle+\beta|10\rangle+\beta|11\rangle)_{5E}. \tag{4.24}$$

If Charlie obtains the state $|1\rangle_2$, then Eve-Bob's system will collapse to $|\Pi^2\rangle_{5E}$, which we denote as

$$|\Pi^2\rangle_{5E}=(\alpha|00\rangle+\alpha|01\rangle+\beta|10\rangle+\beta|11\rangle)_{5E.} \tag{4.25}$$

Therefore, eavesdropper Eve has no chance to get the initial information.

2. Assuming that an eavesdropper exists, after Alice performs the Bell state measurement, the Bob-Charlie-Malicious qubit will collapse into a four-qubit entangled state, as shown below:

$$\begin{aligned}|\xi^1\rangle_{345M} = \frac{1}{4\sqrt{2}}&(a|1010\rangle+a|1011\rangle+a|0100\rangle+a|0101\rangle+b|0000\rangle+b|0001\rangle\\&-b|1110\rangle-b|1111\rangle+c|0010\rangle+c|0011\rangle+c|1100\rangle+c|1101\rangle\\&+d|1000\rangle+d|1001\rangle-d|0110\rangle-d|0111\rangle)_{345M}.\end{aligned} \tag{4.26}$$

If Charlie gets state $|0\rangle_5$, the Malicious-Bob system will collapse to $|\Gamma^1\rangle_{34M}$, say

$$|\Gamma^1\rangle_{34M}=\frac{1}{4\sqrt{2}}(a|010\rangle+a|011\rangle+b|000\rangle+b|001\rangle+c|110\rangle+c|111\rangle+d|100\rangle+d|101\rangle)_{34M}. \tag{4.27}$$

If Charlie gets state $|1\rangle_5$, the Malicious-Bob system will collapse to $|\Gamma^2\rangle_{34M}$, say

$$|\Gamma^2\rangle_{34M} = \frac{1}{4\sqrt{2}}(a|100\rangle + a|101\rangle - b|110\rangle - b|111\rangle + c|000\rangle + c|001\rangle - d|010\rangle - d|011\rangle)_{34M}. \tag{4.28}$$

The Malicious cannot get the initial state information.

Therefore, the quantum information splitting process using five-qubit entangled states to split arbitrary single-qubit and two qubits is safe and reliable.

4.3 Use Four-Qubit Cluster State and Two-Qubit Bell State to Split Arbitrary Two-Qubit State

4.3.1 Separation Process Description

This section describes the splitting of arbitrary two-qubit states by using the four-qubit cluster state and the two-qubit Bell state. It is assumed that Alice, the information sender, has prepared any two qubits $|\Psi\rangle_{AB} = \alpha|00\rangle + \beta|01\rangle + \gamma|10\rangle + \eta|11\rangle$, which satisfy $|\alpha|^2 + |\beta|^2 + |\gamma|^2 + |\eta|^2 = 1$. In general, the coefficients α, β, γ, n are different. If Alice transmits arbitrary two qubits to Bob, the receiver, and Charlie, the controller, then Alice, Bob, and Charlie share a quantum channel, here using the four-qubit cluster state $|\Psi\rangle_{1234} = \frac{1}{2}(|0000\rangle + |0011\rangle + |1100\rangle - |1111\rangle)_{1234}$ and the two-qubit Bell state $|\Phi\rangle_{56} = \frac{1}{\sqrt{2}}(|00\rangle + |11\rangle)_{56}$ as the quantum channel, where Alice has qubits A, B, 1, and 5, Charlie has qubits 2 and 3, and Bob has qubits 4 and 6.

The total quantum state system of qubits owned by Alice, Bob, and Charlie is

$$|\Psi\rangle_{AB123456} = |\Psi\rangle_{AB} \otimes |\Psi\rangle_{1234} \otimes |\Phi\rangle_{56}. \tag{4.29}$$

To complete the quantum information splitting, Alice performs Bell state measurements on the qubit pairs $(A, 1)$ and $(B, 5)$ that she has. After Alice performs the Bell state measurements, Bob and Charlie's qubits will collapse, resulting in 16 possible outcomes $|\phi_{2346}\rangle^i (i = 1, 2, 3, \ldots\ldots, 16)$, as follows:

$$
\begin{aligned}
|\Phi^{++}\rangle_{A1}|\Phi^{\pm}\rangle_{B5} &= |\phi_{2346}\rangle^{1,2} \\
&= \frac{1}{4}(\alpha|0000\rangle + +\alpha|0110\rangle + -\beta|0001\rangle + -\beta|0111\rangle \\
&\quad + +\gamma|1000\rangle - -\gamma|1110\rangle + -\eta|1001\rangle - +\eta|1111\rangle), \\
|\Phi^{++}\rangle_{A1}|\Psi^{\pm}\rangle_{B5} &= |\phi_{2346}\rangle^{3,4} \\
&= \frac{1}{4}(\alpha|0001\rangle + +\alpha|0111\rangle + -\beta|0000\rangle + -\beta|0110\rangle \\
&\quad + +\gamma|1001\rangle - -\gamma|1111\rangle + -\eta|1000\rangle - +\eta|1110\rangle), \\
|\Phi^{--}\rangle_{A1}|\Phi^{\pm}\rangle_{B5} &= |\phi_{2346}\rangle^{5,6} \\
&= \frac{1}{4}(\alpha|0000\rangle + +\alpha|0110\rangle + -\beta|0001\rangle + -\beta|0111\rangle \\
&\quad - -\gamma|1000\rangle + +\gamma|1110\rangle - +\eta|1001\rangle + -\eta|1111\rangle), \\
|\Phi^{--}\rangle_{A1}|\Psi^{\pm}\rangle_{B5} &= |\phi_{2346}\rangle^{7,8} \\
&= \frac{1}{4}(\alpha|0001\rangle + +\alpha|0111\rangle + -\beta|0000\rangle + -\beta|0110\rangle \\
&\quad - -\gamma|1001\rangle + +\gamma|1111\rangle - +\eta|1000\rangle + -\eta|1110\rangle), \\
|\Psi^{++}\rangle_{A1}|\Phi^{\pm}\rangle_{B5} &= |\phi_{2346}\rangle^{9,10} \\
&= \frac{1}{4}(\alpha|1000\rangle - -\alpha|1110\rangle + -\beta|1001\rangle - +\beta|1111\rangle \\
&\quad + +\gamma|0000\rangle + +\gamma|0110\rangle + -\eta|0001\rangle + -\eta|0111\rangle), \\
|\Psi^{++}\rangle_{A1}|\Psi^{\pm}\rangle_{B5} &= |\phi_{2346}\rangle^{11,12} \\
&= \frac{1}{4}(\alpha|1001\rangle - -\alpha|1111\rangle + -\beta|1000\rangle - +\beta|1110\rangle \\
&\quad + +\gamma|0001\rangle + +\gamma|0111\rangle + -\eta|0000\rangle + -\eta|0110\rangle), \\
|\Psi^{--}\rangle_{A1}|\Phi^{\pm}\rangle_{B5} &= |\phi_{2346}\rangle^{13,14} \\
&= \frac{1}{4}\alpha|1000\rangle - -\alpha|1110\rangle + -\beta|1001\rangle - +\beta|1111\rangle.
\end{aligned}
\tag{4.30}
$$

Alice tells the Bell state measurements to Bob and Charlie through the classical channel, and Bob and Charlie must cooperate to recover any two-qubit information, so Charlie performs the Bell state measurement on the qubit (2,3) he owns, and finally Bob reconstructs the unknown quantum state by performing the corresponding unitary operation.

The four unitary transformations performed by Bob are described as

$$
\begin{aligned}
I &= |0\rangle\langle 0| + |1\rangle\langle 1|, & \sigma_x &= |0\rangle\langle 1| + |1\rangle\langle 0|, \\
-i\sigma_y &= |1\rangle\langle 0| - |0\rangle\langle 1|, & \sigma_z &= |0\rangle\langle 0| - |1\rangle\langle 1|.
\end{aligned}
\tag{4.31}
$$

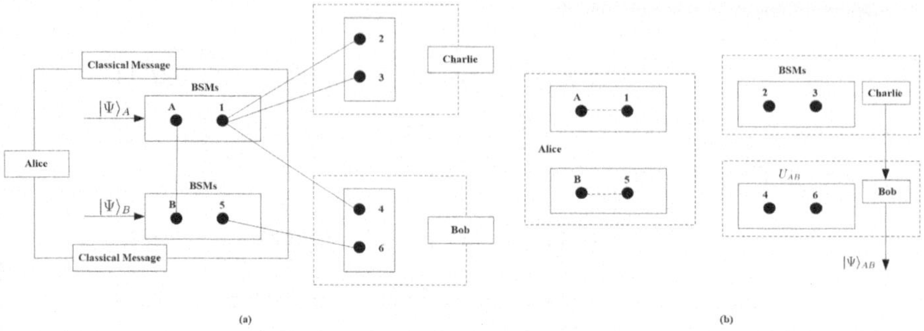

Fig. 4.3 Quantum information splitting process of the arbitrary two qubits

After Charlie performs the Bell state measurement, the qubit owned by Bob collapses instantaneously and 64 possible collapse states are obtained. The collapse state formula $|\phi_{2346}\rangle^{1}$ of Bob and Charlie's qubits was taken as an example for analysis.

1. If Charlie's measurement result is $|00\rangle + |11\rangle$, the first possible collapsed state obtained by Bob is $|\varphi^{1}\rangle_{46} = \alpha|00\rangle + \beta|01\rangle - \gamma|10\rangle - \eta|11\rangle$. Therefore, Bob can recover the initial state information by executing the unitary transformation $\sigma_x \otimes I$.
2. If Charlie's measurement result is $|00\rangle - |11\rangle$, the second possible collapsed state obtained by Bob is $|\varphi^{2}\rangle_{46} = \alpha|00\rangle + \beta|01\rangle + \gamma|10\rangle + \eta|11\rangle$. Therefore, Bob can recover the initial state information by executing the unitary transformation $I \otimes I$.
3. If Charlie's measurement result is $|01\rangle + |10\rangle$, the third possible collapsed state obtained by Bob is $|\varphi^{3}\rangle_{46} = \alpha|10\rangle + \beta|11\rangle + \gamma|00\rangle + \eta|01\rangle$. Therefore, Bob can recover the initial state information by executing the unitary transformation $\sigma_x \otimes I$.
4. If Charlie's measurement result is $|01\rangle - |10\rangle$, the fourth possible collapsed state obtained by Bob is $|\varphi^{4}\rangle_{46} = \alpha|10\rangle + \beta|11\rangle - \gamma|00\rangle - \eta|01\rangle$. Therefore, Bob can recover the initial state information by executing the unitary transformation $-i\sigma_y \otimes I$.

The process of splitting any two qubits using the four-qubit cluster state and two-qubit Bell state is shown in Fig. 4.3.

4.3.2 Security Analysis

Suppose there exists an attacker Eve who tries to steal the communication information without the knowledge of the communicating parties. First, if Eve prepares an auxiliary qubit entangled onto the quantum channel, when Alice finishes the Bell state measurement, the qubit states owned by Bob, Charlie, and Eve will collapse into a five-qubit entangled state, which we can describe as

$$\begin{aligned}|\Xi\rangle 2346E = \frac{1}{4\sqrt{2}}(&\alpha|00000\rangle + \alpha|00001\rangle + |01100\rangle + \alpha|01101\rangle \\ &+\beta|00010\rangle + \beta|00011\rangle + \beta|01110\rangle + \beta|01111\rangle \\ &+\gamma|10000\rangle + \gamma|10001\rangle - \gamma|11100\rangle - \gamma|11101\rangle \\ &+\eta|10010\rangle + \eta|10011\rangle - \eta|11110\rangle - \eta|11111\rangle)_{2346E}.\end{aligned} \tag{4.32}$$

If Charlie obtains the state $|\Phi^{+}\rangle_{23}$, then Bob-Eve's system will collapse to

$$\begin{aligned}|\Pi\rangle_{46E} = &\alpha|000\rangle + \alpha|001\rangle + \beta|010\rangle + \beta|011\rangle - \gamma|100\rangle - \gamma|101\rangle - \eta|110\rangle \\ &- \eta|111\rangle = (\alpha|00\rangle + \beta|01\rangle - \gamma|10\rangle - \eta|11\rangle)_{46}\frac{1}{\sqrt{2}}(|0\rangle + |1\rangle)_{E}.\end{aligned} \tag{4.33}$$

From the above calculation, Eve has no opportunity to obtain the original state information. However, by only considering Eve as the eavesdropper, this only proves that Eve did not steal the original state information during information transmission. However, in the actual transmission process, there may be two eavesdroppers, one of which is Mallory, the malicious attacker, and the other is Eve. The security analysis of quantum teleportation is as follows:

Suppose Mallory and Eve steal the information of the original state by measuring the auxiliary qubits, and attack Alice, Bob, and Charlie in order to steal the information of the original state, if none of the three communicating parties are aware of the existence of the eavesdropper. Then, first Alice performs Bell state measurement; Bob, Charlie, Mallory, and Eve have the qubit state changed instantly and the qubits will collapse to a six-qubit entangled state; then Charlie finishes his Bell state measurement; Bob, Mallory, and Eve's qubits will collapse to a straight product state; Bob passes $|00\rangle \pm |11\rangle$ to reconstruct the original state information. It is obvious that Mallory and Eve do not have any chance to obtain the original state information. This quantum information splitting process is safe for the case that there are two attackers in the communication process.

4.3.3 Physical Implementation

In order to realize the purpose of quantum teleportation, cavity QED is used. First, to achieve quantum information splitting in QED, we must consider the case of multi-qubit entanglement, and we can also use n-qubit. Compared with GHZ states, cluster states have the greatest connectivity and persistent entanglement.

4.3.3.1 Preparation of Cluster States

Consider n two-energy atoms interacting simultaneously with a single-mode cavity field, and in the spin-wave approximation, the system's Hamilton quantity is

$$H = \sum_{i=1}^{N} g_j \left(a^+ |g_j\rangle\langle e_j| + a|e_j\rangle\langle g_j|\right), \tag{4.34}$$

where $|g_i\rangle$ is the excited state of the *j*th atom, g_j is the constant of the mutual coupling between the *j*th atom and the cavity mode, and $a^+(a)$ is the cavity mode creation and annihilates operator after the interaction of the atom and the cavity.

Assuming that the initial state of the cavity field is $|0\rangle_c$, and the initial state of the atom is $|e_1\rangle\prod_{k=2}^{N}|g_k\rangle$, then the evolution of the system over time is shown as follows:

$$\begin{aligned} |e_1\rangle \prod_{k=2}^{N} |g_k\rangle \Big| 0\rangle c &\longrightarrow \left[\left(g_1^2/G^2\right)\cos(Gt) + \left(G^2 - g_1^2\right)/G^2\right]|e_1\rangle \prod_{k=2}^{N}|g_k\rangle \Big| 0\rangle_c \\ &+ \left(g_1/G^2\right)[\cos(Gt) - 1]|g_1\rangle \sum_{k=2}^{N} g_k|e_k\rangle \prod_{i=2,j}^{N}|g_j\rangle \Big| 0\rangle_c - (ig_1/G)\sin(Gt)\prod_{k=1}^{N}|g_k\rangle \Big| 1\rangle_c. \end{aligned} \tag{4.35}$$

When the action time $Gt = \pi$, the state will evolve as.

$$|e_1\rangle \prod_{k=2}^{N} |g_k\rangle\, |0\rangle_c \to 1/\sqrt{2(N-1)}\left(\sqrt{N-1}|e_1\rangle\right) \prod_{k=2}^{N}|g_k\rangle - |g_1\rangle \sum_{k=2}^{N}|e_k\rangle \prod_{j=2,j|g_i\rangle)\,|0\rangle}^{N}. \tag{4.36}$$

Implementing a single quantum bit rotation $|e_1\rangle \to -|e_1\rangle$, the resulting state is

$$\sqrt{2(N-1)\left(\sqrt{N-1}|e_1\rangle \prod_{k=2}^{N}|g_k\rangle\right) + |g_1\rangle \prod_{k=2}^{N}|e_k\rangle \prod_{j=2,j|g_j\rangle)}^{N}}. \tag{4.37}$$

In this way, the entanglement cluster state of *N* qubits can be obtained. If the influence of cavity mode on the cluster state is taken into account, the System Milhadonian can be written as

$$H = \sum_{i=1}^{N} g_{jc}\left(a^+ s_j^- + a s_j^+\right) - \frac{ik}{2a^+ a}, \tag{4.38}$$

where *k* is cavity loss.

Considering the influence of cavity loss on cluster state preparation, the curves of fidelity and success rate with cavity loss are shown in Figs. 4.4 and 4.5.

Through the resonance of multiatom cavity and single-mode cavity, the coupling system of different atoms and cavity modes and the corresponding time are selected to obtain the desired state. Because the resonance model is adopted, the time required is relatively less.

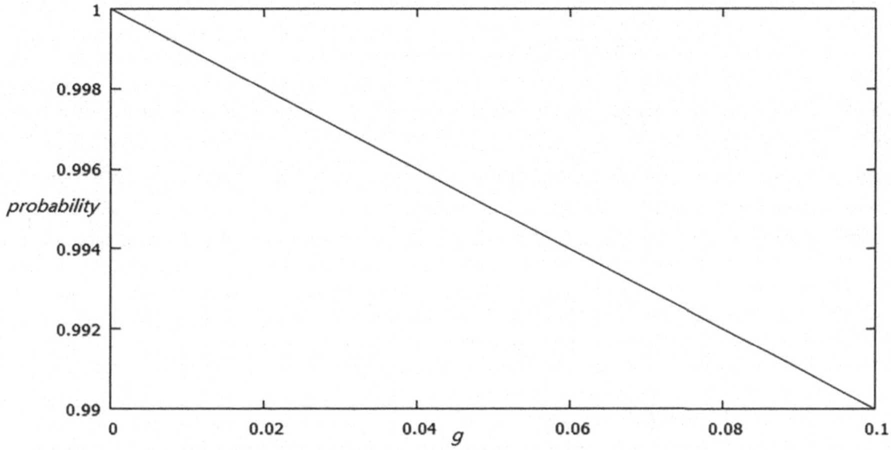

Fig. 4.4 Entanglement preparation with small cavity loss

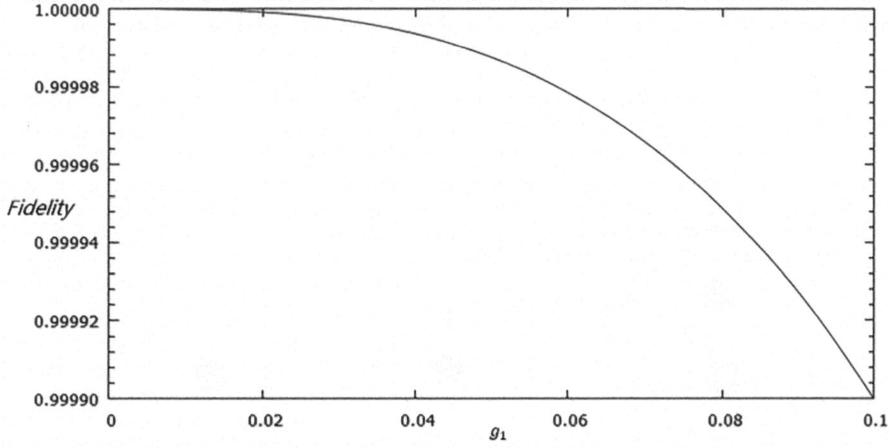

Fig. 4.5 Fidelity changes with cavity loss

4.3.3.2 Preparation of Bell States

In this section, the preparation of Bell states is achieved through bilateral optical microcavities, which are like polarization beam splitters and have different effects on different polarized photons.

By projecting and reflecting photons, the state produced by the system is divided into two parts, part projected and part reflected.

The cavity field emission operator is

$$\hat{t} = |R\rangle\langle R| \otimes |\uparrow\rangle\langle\uparrow| + |L\rangle\langle L| \otimes |\downarrow\rangle\langle\downarrow|, \tag{4.39}$$

where R denotes the right polarization state of the photon, $|\downarrow\rangle$ the down-spin polarization state of the photon in the cavity, $|\uparrow\rangle$ the up-spin polarization state of the photon in the cavity, and L the left polarization state of the photon. The photon and cavity coupling system is shown in Fig. 4.6.

For incoming photons in any state, the reflection operator of the cavity is

$$\begin{aligned}\hat{t} = {} & t_0(w)(|R\rangle\langle R| \otimes |\uparrow\rangle\langle\uparrow| + |L\rangle\langle L| \otimes |\downarrow\rangle\langle\downarrow| \\ & + t_w(R)(|R\rangle\langle R| \otimes |\downarrow\rangle\langle\downarrow| + |L\rangle\langle L| \otimes |\uparrow\rangle\langle\uparrow|.\end{aligned} \tag{4.40}$$

The cavity's projection operator is

$$\begin{aligned}\hat{r} = {} & r_0(w) \\ & \times (|R\rangle\langle R| \otimes |\uparrow\rangle\langle\uparrow| + |L\rangle\langle L| \otimes |\downarrow\rangle\langle\downarrow| + r_w(R)(|R\rangle\langle R| \otimes |\downarrow\rangle\langle\downarrow| + |L\rangle\langle L| \otimes |\uparrow\rangle\langle\uparrow|.\end{aligned}$$

For photon A in any state shooting into the cavity field, the entangled state formed by the interaction between the incoming photon and the atom in the cavity is

$$|\psi^e\rangle \otimes |\psi^s\rangle\overrightarrow{\hat{t}(w)}t_0(w)/\sqrt{2}(\alpha|R\rangle|\uparrow\rangle + \beta|L\rangle|\downarrow\rangle)\,. \tag{4.41}$$

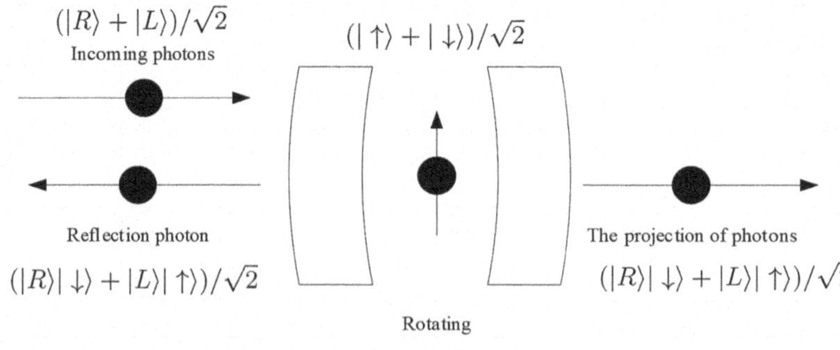

Fig. 4.6 Evolution process of photonic cavity coupling system

The projected photon performs an H-gate operation and uses the nondestructive measurement principle to measure the state of the electron, thus realizing the transfer of an electron to the photon.

When photons 1 and 2 enter the cavity field, the state of the atoms in the cavity is $|\psi^s\rangle = (|\uparrow\rangle + |\downarrow\rangle)\sqrt{2}$ due to the fact that the two photons have the same superposition state $|\psi^{ph}\rangle_{1,2} = (\alpha|R\rangle_{1,2} + \beta|L\rangle_{1,2})\sqrt{2}$. Then, the incoming photons interact with the atoms in the cavity, producing the following results:

1. If the photons are all projected out of the cavity field, the evolution of the entangled state is

$$|\psi^{ph}\rangle \otimes |\psi^{ph}\rangle \otimes |\psi\rangle^s \rightarrow (t_0^2(w))/2(|R\rangle_1|R\rangle_2|\uparrow\rangle + |L\rangle_1|L\rangle_2|\downarrow\rangle)/\sqrt{2}. \tag{4.42}$$

   Perform an H-gate operation on the atoms in the cavity; then the quantum change is

$$|\uparrow\rangle \rightarrow (|\uparrow\rangle + |\downarrow\rangle)/\sqrt{2}, \quad |\downarrow\rangle \rightarrow (|\uparrow\rangle - |\downarrow\rangle)/\sqrt{2}. \tag{4.43}$$

   When the photon is projected out, the state at the projection port can be transformed into

$$|\psi^{ph}\rangle \otimes |\psi^{ph}\rangle \otimes |\psi\rangle^s \rightarrow \frac{1}{2}(|R\rangle_1|R\rangle_2 + |L\rangle_1|L\rangle_2|\uparrow\rangle + \frac{1}{2}(|R\rangle_1|R\rangle_2 - |L\rangle_1|L\rangle_2|\downarrow\rangle. \tag{4.44}$$

   Then we measure the state of the electron to get the state of the photon pair, and finally we get the Bell state:

$$|\Psi^{\pm}\rangle_{1,2} = (|R\rangle_1|R\rangle_2 \pm |L\rangle_1|L\rangle_2)/\sqrt{2}. \tag{4.45}$$

   This happens 25% of the time, as shown in Fig. 4.7.
2. Similarly, if all the photons are reflected from the cavity field, there is only a 25% chance of that happening, as shown in Fig. 4.8.

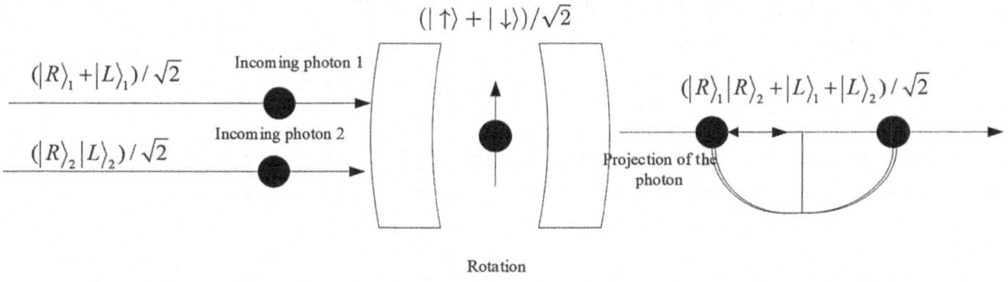

Fig. 4.7 Schematic diagram of all photons being projected

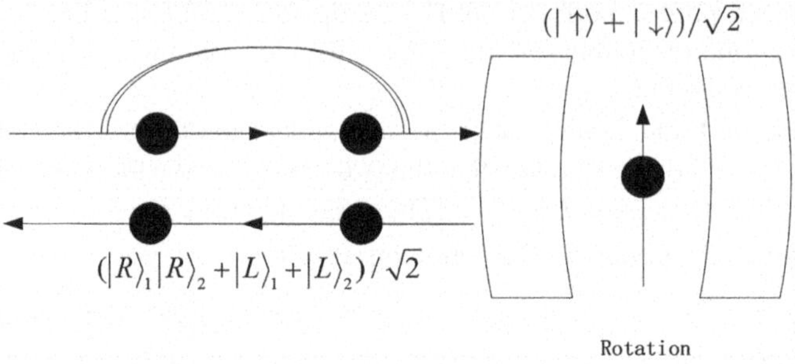

Fig. 4.8 Schematic diagram of all photons reflected out

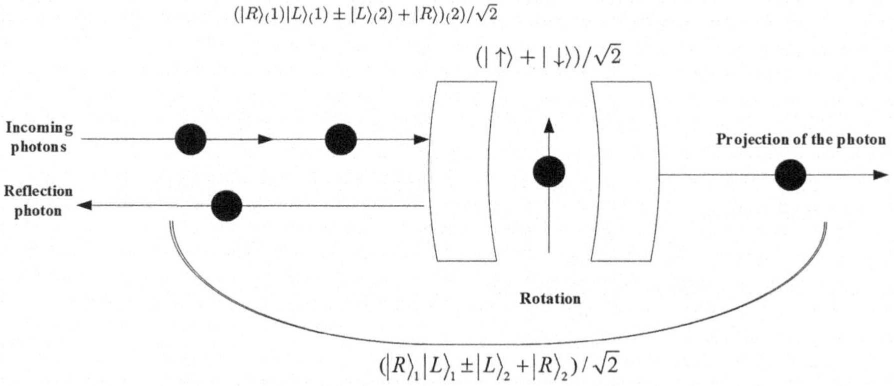

Fig. 4.9 Interaction between photon and cavity

3. If two photons enter the cavity field and one photon is reflected out while the other photon is projected after passing through the cavity field, the probability of this happening is 50%. After the photon interacts with the atom in the cavity field, the quantum state evolves as

$$\left|\psi^{ph}\right\rangle \otimes \left|\psi^{ph}\right\rangle \otimes |\psi\rangle^{s} \rightarrow (t_0(w)r(w))/2(|R\rangle_1|L\rangle_2|\uparrow\rangle+|L\rangle_1|R\rangle_2|\downarrow\rangle)/\sqrt{2}. \quad (4.46)$$

With the same method as 1, we get the Bell state as

$$|\Phi^{\pm}\rangle_{1,2} = (|R\rangle_1|R\rangle_2 \pm |L\rangle_1|L\rangle_2)/\sqrt{2}. \quad (4.47)$$

The interaction process between photon and cavity field is shown in Fig. 4.9.

4.3.3.3 The Process Realized in QED System

In this scenario, Alice wants to transmit the arbitrary two-qubit state; Alice, Bob, and Charlie share a quantum entanglement channel formed by a four-qubit cluster state and a two-qubit Bell state; and the QED system is used to describe the process of this scenario.

Considering six two-energy atoms interacting simultaneously with a unimodal cavity and a classical field, in the spin-wave approximation, we describe the Milharton quantity of the system as

$$H = w_0 \sum_{j=1}^{6} S_{z,j} + w_a a^+ a + \sum_{j=1}^{6} \left[g\left(a^+ S_j^- + a S_j^+\right) + \Theta\left(S_j^+ \exp(-iwt) + S_j^- \exp(iwt)\right)\right], \tag{4.48}$$

where w_0 is the atomic transition frequency, w_a is the cavity field frequency, w is the driving field frequency, a^+ is the operator for cavity mode generation, a is the operator for cavity mode annihilation, g is the coefficient of cavity-field-atom coupling, and Θ is the Rabi frequency.

Suppose $w_0 = w$; the Milharton quantity of the system after the interaction is

$$H_i = \sum_{i=1}^{6} g\left(e^{-i\delta t} a^+ s_j^- + e^{i\delta t} a s_j^+\right) + \sum_{i=1}^{6} \Theta\left(s_j^+ + s_j^-\right). \tag{4.49}$$

When $\lambda = g^2/2\delta$, the system operator evolves into

$$U(t) = e^{-iH_0 t} e^{-iH_e t}, \tag{4.50}$$

where $H_0 = \sum_{i=1}^{6} \Theta\left(s_j^+ + s_j^-\right) t$; if Alice wants to separate arbitrary two-qubit states, then she first performs the Bell state measurement in QED and applies a rotation $|1\rangle \rightarrow i \mid 1\rangle$ to the qubit, which will give the following result:

$$\begin{aligned} |\Phi^+\rangle_{ij} &\rightarrow \frac{1}{\sqrt{2}}(|00\rangle + i|11\rangle)_{ij}, \quad |\Phi^-\rangle_{ij} \rightarrow \frac{1}{\sqrt{2}}(|00\rangle - i|11\rangle)_{ij}, \\ |\Psi^+\rangle_{ij} &\rightarrow \frac{1}{\sqrt{2}}(|01\rangle + i|10\rangle)_{ij}, \quad |\Psi^-\rangle_{ij} \rightarrow \frac{1}{\sqrt{2}}(|01\rangle - i|10\rangle)_{ij}. \end{aligned} \tag{4.51}$$

Let qubit i and qubit j interact in a single-mode cavity and drive in a classical strong field. Choose the time and Rabi frequency of interaction to make $\lambda t = \frac{\pi}{4}$. In this way, we get the evolutionary state as follows:

$$
\begin{aligned}
&|\Phi^+\rangle_{ij} \rightarrow |00\rangle_{ij}, \quad |\Phi^-\rangle_{ij} \rightarrow -i|11\rangle_{ij}, \\
&|\Psi^+\rangle_{ij} \rightarrow \Big|01\rangle_{ij}, \quad |\Psi^-\rangle_{ij} \rightarrow -i\Big|10\rangle_{ij}.
\end{aligned} \tag{4.52}
$$

Discarding the phase factor $e^{-i\pi/4}$, we complete the Bell state measurement by measuring particle i and particle j separately, so that Bob can successfully reconstruct the information of the original state $|\Psi\rangle_{AB}$.

4.3.4 Use Four-Qubit Cluster State and GHZ State to Split Arbitrary Three-Qubit State

4.3.5 Description of the Splitting Process

The three communicating parties, Alice, Bob, and Charlie, share a seven-qubit quantum channel consisting of a four-qubit cluster state $|\Psi\rangle_{1234} = \frac{1}{2} \times (|0000\rangle + |0011\rangle + |1100\rangle - |1111\rangle)$ and a three-qubit GHZ state $|\mathrm{GHZ}\rangle_{567} = \frac{1}{\sqrt{2}}(|000\rangle + |111\rangle)$.

Step 1: Alice wants to send the unknown three-qubit state $|\Psi\rangle_{ABC} = a|000\rangle + b|001\rangle + c|010\rangle + d|011\rangle + e|100\rangle + f|101\rangle + g|110\rangle + h|111\rangle$, $|001\rangle + c|010\rangle + d|011\rangle + e|100\rangle + f|101\rangle + g|110\rangle + h|111\rangle$, satisfying $|a|^2 + |b|^2 + |c|^2 + |d|^2 + |e|^2 + |f|^2 + |g|^2 + |h|^2 = 1$, to the recipients Bob and Charlie. Alice has qubits A, B, C, 1, 3 and 5; Bob has qubits 2, 4, and 6; and Charlie has qubit 7. The total combined state of the whole system is

$$
|\Psi\rangle_{ABC1234567} = |\Psi\rangle_{ABC} \otimes |\Psi\rangle_{1234} \otimes |\mathrm{GHZ}\rangle_{567}, \tag{4.53}
$$

where

$$
\begin{aligned}
\Psi\rangle_{1234567} = |\Psi\rangle_{1234} \otimes |\mathrm{GHZ}\rangle_{567} \\
= \frac{1}{2\sqrt{2}}(|0000000\rangle + |0011000\rangle + |1100000\rangle \\
- |1111000\rangle + |0000111\rangle + |0011111\rangle + |1100111\rangle - |1111111\rangle).
\end{aligned}
$$

We describe it in detail as

$$
\begin{aligned}
|\Psi\rangle ABC1234567 = {} & |\Psi\rangle_{ABC} \otimes |\Psi\rangle_{1234567} \\
= {} & \frac{1}{2\sqrt{2}}\{(a|00000000001\rangle + a|0000011000\rangle \\
& + a|0001100000\rangle - a|0001111000\rangle \\
& + a|0000000111\rangle + a|0000011111\rangle \\
& + a|0001100111\rangle - a|0001111111\rangle) \\
& + (b|0010000000\rangle + b|0010011000\rangle \\
& + b|0011100000\rangle - b|001111100\rangle \\
& + b|0010000111\rangle + b|0010011111\rangle \\
& + b|0011100111\rangle - b|0011111111\rangle) \\
& + (c|0100000000\rangle + c|0100011000\rangle \\
& + c|0101100000\rangle - c|0101111000\rangle \\
& + c|0100000111\rangle + c|0100011111\rangle \\
& + c|0101100111\rangle - c|0101111111\rangle) \\
& + (d|0110000000\rangle + d|0110011000\rangle \\
& + d|0111100000\rangle - d|0111111000\rangle \\
& + d|0110000111\rangle + d|0110011111\rangle \\
& + d|0111100111\rangle - d|0111111111\rangle) \\
& + (e|1000000000\rangle + e|1000011000\rangle \\
& + e|1001100000\rangle - e|1001111000\rangle \\
& + e|1000000111\rangle + e|1000011111\rangle \\
& + e|1001100111\rangle - e|1001111111\rangle) \\
& + (f|1010000000\rangle + f|1010011000\rangle \\
& + f|1011100000\rangle - f|1011111000\rangle \\
& + f|1010000111\rangle + f|1010011111\rangle \\
& + f|1011100111\rangle - f|1011111111\rangle) \\
& + (g|1100000000\rangle + g|1100011000\rangle \\
& + g|1101100000\rangle - g|1101111000\rangle \\
& + g|1100000111\rangle + g|1100011111\rangle \\
& + g|1101100111\rangle - g|1101111111\rangle) \\
& + (h|1110000000\rangle + h|1110011000\rangle \\
& + h|1111100000\rangle - h|1111111000\rangle \\
& + h|1110000111\rangle + h|1110011111\rangle \\
& + h|1111100111\rangle - h|1111111111\rangle)\}.
\end{aligned}
\tag{4.54}
$$

Step 2: In order to split the original state information and send the information to the information receiver Bob, Alice first performs Bell state measurement on qubit pairs (A, 1), (B, 3), and (C, 5). After Alice completes the measurement, the qubits owned by Bob and Charlie collapse, and the collapse state is one of the following 64 states.

$$|\Phi^{++}\rangle_{A1}|\Phi^{++}\rangle_{B3}|\Phi^{+-}\rangle_{C5}=\frac{1}{8}(a|0000\rangle+-b|0011\rangle++c|0100\rangle+-d|0111\rangle$$
$$++e|1000\rangle+-f|1011\rangle--g|1100\rangle-+h|1111\rangle),$$

$$|\Phi^{++}\rangle_{A1}|\Phi^{++}\rangle_{B3}|\Psi^{+-}\rangle_{C5}=\frac{1}{8}(a|0011\rangle+-b|0000\rangle++c|0111\rangle+-d|0100\rangle$$
$$++e|1011\rangle+-f|1000\rangle-+g|1111\rangle--h|1100\rangle),$$

$$|\Phi^{++}\rangle_{A1}|\Phi^{--}\rangle_{B3}|\Phi^{+-}\rangle_{C5}=\frac{1}{8}(a|0000\rangle+-b|0011\rangle--c|0100\rangle-+d|0111\rangle$$
$$++e|1000\rangle+-f|1011\rangle++g|1100\rangle+-h|1111\rangle),$$

$$|\Phi^{++}\rangle_{A1}|\Phi^{--}\rangle_{B3}|\Psi^{+-}\rangle_{C5}=\frac{1}{8}(a|0011\rangle+-b|0000\rangle--c|0111\rangle-+d|0100\rangle$$
$$++e|1011\rangle+-f|1000\rangle++g|1111\rangle+-h|1100\rangle),$$

$$|\Phi^{--}\rangle_{A1}|\Phi^{++}\rangle_{B3}|\Phi^{+-}\rangle_{C5}=\frac{1}{8}(a|0000\rangle+-b|0011\rangle++c|0100\rangle+-d|0111\rangle$$
$$--e|1000\rangle-+f|1011\rangle++g|1100\rangle+-h|1111\rangle),$$

$$|\Phi^{--}\rangle_{A1}|\Phi^{++}\rangle_{B3}|\Psi^{+-}\rangle_{C5}=\frac{1}{8}(a|0011\rangle+-b|0000\rangle++c|0111\rangle+-d|0100\rangle$$
$$--e|1011\rangle-+f|1000\rangle++g|1111\rangle+-h|1100\rangle),$$

$$|\Phi^{--}\rangle_{A1}|\Phi^{--}\rangle_{B3}|\Phi^{+-}\rangle_{C5}=\frac{1}{8}(a|0000\rangle+-b|0011\rangle--c|0100\rangle-+d|0111\rangle$$
$$--e|1000\rangle-+f|1011\rangle--g|1100\rangle-+h|1111\rangle),$$

$$|\Phi^{--}\rangle_{A1}|\Phi^{--}\rangle_{B3}|\Psi^{+-}\rangle_{C5}=\frac{1}{8}(a|0011\rangle+-b|0000\rangle--c|0111\rangle-+d|0100\rangle$$
$$--e|1011\rangle-+f|1000\rangle--g|1111\rangle-+h|1100\rangle),$$

$$|\Phi^{++}\rangle_{A1}|\Psi^{++}\rangle_{B3}|\Phi^{+-}\rangle_{C5}=\frac{1}{8}(a|0100\rangle+-b|0111\rangle++c|0000\rangle+-d|0011\rangle$$
$$--e|1100\rangle-+f|1111\rangle++g|1000\rangle+-h|1011\rangle),$$

$$|\Phi^{++}\rangle_{A1}|\Psi^{++}\rangle_{B3}|\Psi^{+-}\rangle_{C5}=\frac{1}{8}(a|0111\rangle+-b|0100\rangle++c|0011\rangle+-d|0000\rangle$$
$$--e|1111\rangle-+f|1100\rangle++g|1011\rangle+-h|1000\rangle),$$

$$|\Phi^{++}\rangle_{A1}|\Psi^{--}\rangle_{B3}|\Phi^{+-}\rangle_{C5}=\frac{1}{8}(a|0100\rangle+-b|0111\rangle--c|0000\rangle-+d|0011\rangle$$
$$--e|1100\rangle-+f|1111\rangle--g|1000\rangle-+h|1011\rangle),$$

$$|\Phi^{++}\rangle_{A1}|\Psi^{--}\rangle_{B3}|\Psi^{+-}\rangle_{C5}=\frac{1}{8}(a|0111\rangle+-b|0100\rangle--c|0011\rangle--d|0000\rangle$$
$$--e|1111\rangle-+f|1100\rangle--g|1011\rangle++h|1000\rangle),$$

$$|\Phi^{--}\rangle_{A1}|\Psi^{++}\rangle_{B3}|\Phi^{+-}\rangle_{C5}=\frac{1}{8}(a|0100\rangle+-b|0111\rangle++c|0000\rangle+-d|0011\rangle$$
$$++e|1100\rangle+-f|1111\rangle--g|1000\rangle-+h|1011\rangle),$$

$$|\Phi^{--}\rangle_{A1}|\Psi^{++}\rangle_{B3}|\Psi^{+-}\rangle_{C5}=\frac{1}{8}(a|0111\rangle+-b|0100\rangle++c|0011\rangle+-d|0000\rangle$$
$$++e|1111\rangle+-f|1100\rangle--g|1011\rangle-+h|1000\rangle),$$

$$
\begin{aligned}
|\Phi^{--}\rangle_{A1}|\Psi^{--}\rangle_{B3}|\Phi^{+-}\rangle_{C5} = \frac{1}{8}(a|0100\rangle + -b|0111\rangle - -c|0000\rangle - +d|0011\rangle \\
+ + e|1100\rangle + -f|1111\rangle + +g|1000\rangle + -h|1011\rangle),
\end{aligned}
$$

$$
\begin{aligned}
|\Phi^{--}\rangle_{A1}|\Psi^{--}\rangle_{B3}|\Psi^{+-}\rangle_{C5} = \frac{1}{8}(a|0111\rangle + -b|0100\rangle - -c|0011\rangle - +d|0000\rangle \\
+ + e|1111\rangle + -f|1100\rangle + +g|1011\rangle + -h|1000\rangle),
\end{aligned}
$$

$$
\begin{aligned}
|\Psi^{++}\rangle_{A1}|\Psi^{++}\rangle_{B3}|\Phi^{+-}\rangle_{C5} = \frac{1}{8}(- -a|1100\rangle - +b|1111\rangle + +c|1000\rangle + -d|1011\rangle \\
+ + e|0100\rangle + -f|0111\rangle + +g|0000\rangle + -h|0011\rangle),
\end{aligned}
$$

$$
\begin{aligned}
|\Psi^{++}\rangle_{A1}|\Psi^{++}\rangle_{B3}|\Psi^{+-}\rangle_{C5} = \frac{1}{8}(- -a|1111\rangle - +b|1100\rangle + +c|1011\rangle + -d|1000\rangle \\
+ + e|0111\rangle + -f|0100\rangle + +g|0011\rangle + -h|0000\rangle),
\end{aligned}
$$

$$
\begin{aligned}
|\Psi^{++}\rangle_{A1}|\Psi^{--}\rangle_{B3}|\Phi^{+-}\rangle_{C5} = \frac{1}{8}(- -a|1100\rangle - +b|1111\rangle - -c|1000\rangle - +d|1011\rangle \\
+ + e|0100\rangle + -f|0111\rangle - -g|0000\rangle - +h|0011\rangle),
\end{aligned}
$$

$$
\begin{aligned}
|\Psi^{++}\rangle_{A1}|\Psi^{--}\rangle_{B3}|\Psi^{+-}\rangle_{C5} = \frac{1}{8}(- -a|1111\rangle - +b|1100\rangle - -c|1011\rangle - +d|1000\rangle \\
+ + e|0111\rangle + -f|0100\rangle - -g|0011\rangle - +h|0000\rangle),
\end{aligned}
$$

$$
\begin{aligned}
|\Psi^{--}\rangle_{A1}|\Psi^{++}\rangle_{B3}|\Phi^{+-}\rangle_{C5} = \frac{1}{8}(- -a|1100\rangle - +b|1111\rangle + +c|1000\rangle + -d|1011\rangle \\
- -e|0100\rangle - +f|0111\rangle - 2|0000\rangle - +h|0011\rangle),
\end{aligned}
$$

$$
\begin{aligned}
|\Psi^{--}\rangle_{A1}|\Psi^{++}\rangle_{B3}|\Psi^{+-}\rangle_{C5} = \frac{1}{8}(- -a|1111\rangle - +b|1100\rangle + +c|1011\rangle + -d|1000\rangle \\
- -e|0111\rangle - +f|0100\rangle - -g|0011\rangle - +h|0000\rangle),
\end{aligned}
$$

$$
\begin{aligned}
|\Psi^{--}\rangle_{A1}|\Psi^{--}\rangle_{B3}|\Phi^{+-}\rangle_{C5} = \frac{1}{8}(- -a|1100\rangle - +b|1111\rangle - -c|1000\rangle - +d|1011\rangle \\
- -e|0100\rangle - +f|0111\rangle + +g|0000\rangle + -h|0011\rangle),
\end{aligned}
$$

$$
\begin{aligned}
|\Psi^{--}\rangle_{A1}|\Psi^{--}\rangle_{B3}|\Psi^{+-}\rangle_{C5} = \frac{1}{8}(- -a|1111\rangle - +b|1100\rangle - -c|1011\rangle - +d|1000\rangle \\
- -e|0111\rangle - +f|0100\rangle + +g|0011\rangle + -h|0000\rangle),
\end{aligned}
$$

$$
\begin{aligned}
|\Psi^{++}\rangle_{A1}|\Phi^{++}\rangle_{B3}|\Phi^{+-}\rangle_{C5} = \frac{1}{8}(a|1000\rangle + -b|1011\rangle - -c|1100\rangle - +d|1111\rangle \\
+ + e|0000\rangle + -f|0011\rangle + +g|0100\rangle + -h|0111\rangle),
\end{aligned}
$$

$$
\begin{aligned}
|\Psi^{++}\rangle_{A1}|\Phi^{++}\rangle_{B3}|\Psi^{+-}\rangle_{C5} = \frac{1}{8}(a|1011\rangle + -b|1000\rangle - -c|1111\rangle - +d|1100\rangle \\
+ + e|0011\rangle + -f|0000\rangle + +g|0111\rangle + -h|0100\rangle),
\end{aligned}
$$

$$
\begin{aligned}
|\Phi^{++}\rangle_{A1}|\Psi^{--}\rangle_{B3}|\Phi^{++}\rangle_{C5} = \frac{1}{8}(a|0100\rangle + +b|0111\rangle - -c|0000\rangle - -d|0011\rangle \\
- -e|1100\rangle - -f|1111\rangle - -g|1000\rangle - -h|1011\rangle),
\end{aligned}
$$

$$
\begin{aligned}
|\Phi^{++}\rangle_{A1}|\Psi^{--}\rangle_{B3}|\Psi^{+-}\rangle_{C5} = \frac{1}{8}(a|0111\rangle + -b|0100\rangle - -c|0011\rangle - +d|0000\rangle \\
- -e|1111\rangle - +f|1100\rangle - -g|1011\rangle - +h|1000\rangle),
\end{aligned}
$$

$$
\begin{aligned}
|\Psi^{--}\rangle_{A1}|\Phi^{++}\rangle_{B3}|\Phi^{+-}\rangle_{C5} &= \frac{1}{8}(a|1000\rangle + -b|1011\rangle - -c|1100\rangle - +d|1111\rangle \\
&\quad - -e|0000\rangle - +f|0011\rangle - -g|0100\rangle - +h|0111\rangle), \\
|\Psi^{--}\rangle_{A1}|\Phi^{++}\rangle_{B3}|\Psi^{+-}\rangle_{C5} &= \frac{1}{8}(a|1011\rangle + -b|1000\rangle - +c|1111\rangle - -d|1100\rangle \\
&\quad - -e|0011\rangle - +f|0000\rangle - -g|0111\rangle - +h|0100\rangle), \\
|\Psi^{--}\rangle_{A1}|\Phi^{--}\rangle_{B3}|\Phi^{+-}\rangle_{C5} &= \frac{1}{8}(a|1000\rangle + -b|1011\rangle + +c|1100\rangle + -d|1111\rangle \\
&\quad - -e|0000\rangle - +f|0011\rangle + +g|0100\rangle + -h|0111\rangle), \\
|\Psi^{--}\rangle_{A1}|\Phi^{--}\rangle_{B3}|\Psi^{+-}\rangle_{C5} &= \frac{1}{8}(a|1011\rangle + -b|1000\rangle + +c|1111\rangle + -d|1100\rangle \\
&\quad - -e|0011\rangle - +f|0000\rangle + +g|0111\rangle + -h|0100\rangle).
\end{aligned}
\tag{4.55}
$$

Step 3: Alice then informs Bob and Charlie of the measurements via the classical channel, and if Charlie allows Bob to reconstruct the original state, he must perform a single-qubit measurement on qubit 7 based on $\{|\pm\rangle_7\}$, where $|+\rangle = \frac{1}{\sqrt{2}}(|0\rangle + |1\rangle)$ and $|-\rangle = \frac{1}{\sqrt{2}}(|0\rangle - |1\rangle)$. With the results of Alice and Charlie's measurements, Bob reconstructs the information of the original state $|\Psi\rangle_{ABC}$ after performing the unitary operation (U_1, U_2, U_3, U_4) on his qubits 2, 4, and 6. The unitary operation performed by Bob can be written as follows:

$$
\begin{aligned}
U_1 = {} & |000\rangle\langle 000| + |001\rangle\langle 010| + |010\rangle\langle 100| + |011\rangle\langle 110| + |100\rangle\langle 001| \\
& - |101\rangle\langle 011| - |110\rangle\langle 101| + |111\rangle\langle 110|,
\end{aligned}
\tag{4.56}
$$

$$
\begin{aligned}
U_2 = {} & |000\rangle\langle 000| + |001\rangle\langle 010| + |010\rangle\langle 100| + |011\rangle\langle 110| + |100\rangle\langle 001| \\
& + |101\rangle\langle 011| - |110\rangle\langle 101| - |111\rangle\langle 110|,
\end{aligned}
\tag{4.57}
$$

$$
\begin{aligned}
U_3 = {} & |000\rangle\langle 000| - |001\rangle\langle 010| - |010\rangle\langle 100| - |011\rangle\langle 110| + |100\rangle\langle 001| \\
& + |101\rangle\langle 011| - |110\rangle\langle 101| - |111\rangle\langle 110|,
\end{aligned}
\tag{4.58}
$$

$$
\begin{aligned}
U_4 = {} & |000\rangle\langle 000| - |001\rangle\langle 010| - |010\rangle\langle 100| - |011\rangle\langle 110| - |100\rangle\langle 001| \\
& - |101\rangle\langle 011| + |110\rangle\langle 101| + |111\rangle\langle 110|.
\end{aligned}
\tag{4.59}
$$

Now, take ${}_{A1}\langle\Phi^{+}|_{B3}\langle\Phi^{+}|_{C5}\langle\Phi^{+}|\Psi\rangle_{ABC1234567}$ as an example to illustrate the process of quantum information splitting.

$$
\begin{aligned}
&|\Phi^+\rangle_{A1}|\Phi^+\rangle_{B3}|\Phi^+\rangle_{C5} \\
=&\frac{1}{8}(a|0000\rangle+b|1100\rangle+c|0100\rangle+d|0111\rangle+e|1000\rangle+f|1011\rangle-g|1100\rangle-h|1111\rangle) \\
=&\frac{\sqrt{2}}{2}[|+\rangle_7(a|000\rangle+b|001\rangle+c|010\rangle+d|011\rangle+e|100\rangle+f|101\rangle \\
&-g|110\rangle-h|111\rangle)_{246}]+\frac{\sqrt{2}}{2}[|-\rangle_7(a|000\rangle+b|001\rangle+c|010\rangle+d|011\rangle \\
&+e|100\rangle+f|101\rangle-g|110\rangle-h|111\rangle)_{246}].
\end{aligned}
\tag{4.60}
$$

Charlie performs a single-qubit measurement on his qubit 7 based on $\{|\pm\rangle_7\}$ and tells Bob the result. If the single-qubit measurement is $|+\rangle$ or $|-\rangle$, the remaining qubits will collapse into one of the following states:

$$
|\varphi^1\rangle=(a|000\rangle+b|001\rangle+c|010\rangle+d|011\rangle+e|100\rangle+f|101\rangle-g|110\rangle-h|111\rangle)_{246}
\tag{4.61}
$$

or

$$
|\varphi^2\rangle=(a|000\rangle+b|001\rangle+c|010\rangle+d|011\rangle+e|100\rangle-f|101\rangle-g|110\rangle+h|111\rangle)246.
\tag{4.62}
$$

To obtain the original state information, Bob must perform the single-qubit unitary operation on his qubit 3, and the converted state is as follows:

$$
\begin{aligned}
&a|000\rangle+b|001\rangle+c|010\rangle+d|011\rangle+e|100\rangle+f|101\rangle+g|110\rangle \\
&+h|111\rangle.
\end{aligned}
\tag{4.63}
$$

If the single-qubit measurement results in $|+\rangle$, Bob must perform unitary operation U_2 on his qubits 2, 4, and 6 to reconstruct the unknown original state information.

If the single-qubit measurement results in $|-\rangle$, Bob must perform unitary operation U_1 on his qubits 2, 4, and 6 to reconstruct the unknown original state information.

Figure 4.10 shows the quantum information splitting process.

4.3.6 Security Analysis

In the scheme, the four-qubit cluster state and Bell state are used to realize the quantum information splitting of any three qubits. It is assumed that there is no attacker and the communication is safe and reliable. If there is an attacker, the analysis is as follows:

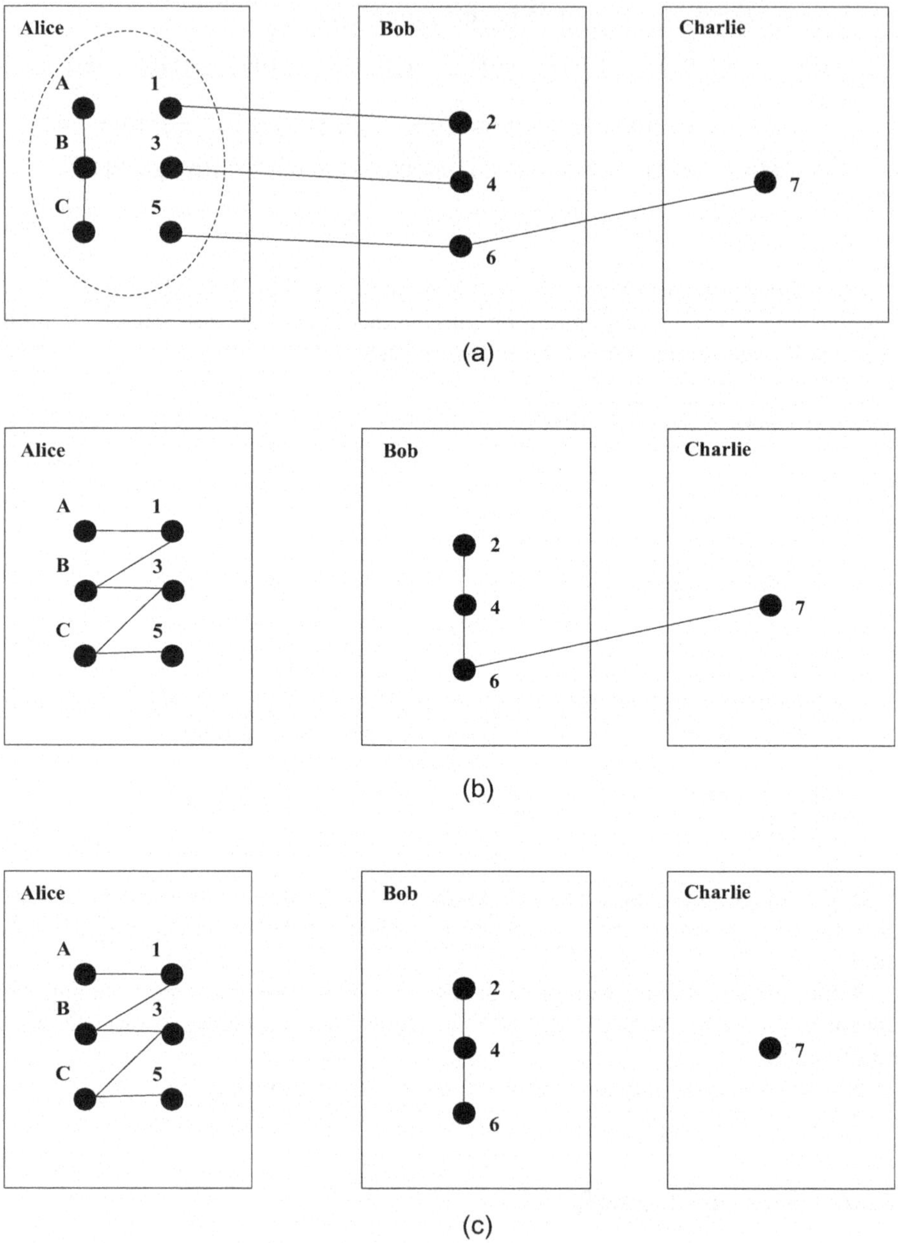

Fig. 4.10 Subgraphs (**a**), (**b**), and (**c**) show the quantum information splitting process

Assuming that the attacker is Eve, who tries to steal the information sent by the three communicating parties without their awareness, he first steals the original state information by measuring the entangled auxiliary qubits, and after Alice performs the joint Bell state measurement, the qubits of Bob, Charlie, and Eve will collapse to a five-qubit entangled state. Charlie then performs a single-qubit measurement on his qubit 7, and finally the Bob-Eve system will collapse to a straight product state, so that Eve will not have any chance to obtain the original unknown state information.

The entangled qubit attached to qubit 7 is $\frac{1}{\sqrt{2}}(|0\rangle+|1\rangle)$. If the result obtained by Alice is ${}_{A1}|\Phi^{+}\rangle_{B3}|\Phi^{+}\rangle_{C5}|\Phi^{+}\rangle$, then the joint state of Bob, Charlie, and Eve is

$$\begin{aligned}|\Pi\rangle 2467E = & \frac{1}{8\sqrt{2}}(a|00000\rangle + a|00001\rangle + b|00110\rangle + b|00111\rangle \\ & + c|01000\rangle + c|01001\rangle + d|01110\rangle + d|01111\rangle + e|10000\rangle + e|10001\rangle \\ & + f|10110\rangle + f|10111\rangle - g|11000\rangle - g|11001\rangle - h|11110\rangle \\ & - h|11111\rangle)_{2467E}.\end{aligned} \tag{4.64}$$

If Charlie obtains state $\{|+\rangle_7\}$, then Bob-Eve's system will collapse to state $|\xi_1\rangle_{246E}$:

$$\begin{aligned}|\xi_1\rangle_{246E} = & (a|0000\rangle + a|0001\rangle + b|0010\rangle + b|0011\rangle + c|0100\rangle \\ & + c|0101\rangle + d|0110\rangle + d|0111\rangle + e|1000\rangle + e|1001\rangle \\ & + f|1010\rangle + f|1011\rangle - g|1100\rangle - g|1101\rangle - h|1110\rangle - h|1111\rangle)_{246E} \\ = & (a|000\rangle + b|001\rangle + c|010\rangle + d|011\rangle + e|100\rangle + f|101\rangle - g|110\rangle \\ & - h|111\rangle)_{246} \frac{1}{\sqrt{2}}(|0\rangle + |1\rangle)_E.\end{aligned} \tag{4.65}$$

If Charlie obtains state $\{|-\rangle_7\}$, then Bob-Eve's system will collapse to state $|\xi_2\rangle_{246E}$:

$$\begin{aligned}|\xi_2\rangle_{246E} = & (a|0000\rangle + a|0001\rangle + b|0010\rangle + b|0011\rangle + c|0100\rangle + c|0101\rangle + d|0110\rangle \\ & + d|0111\rangle + e|1000\rangle + e|1001\rangle - f|1010\rangle - f|1011\rangle - g|1100\rangle - g|1101\rangle \\ & + h|1110\rangle + h|1111\rangle)_{246E} \\ = & (a|000\rangle + b|001\rangle + c|010\rangle + d|011\rangle + e|100\rangle - f|101\rangle - g|110\rangle \\ & + h|111\rangle)_{246} \frac{1}{\sqrt{2}}(|0\rangle + |1\rangle)_E.\end{aligned} \tag{4.66}$$

Obviously, Eve doesn't have any access to the original information.

4.3.7 Physical Implementation

If the spin electron is in state $|\uparrow\rangle$, the incoming photon is in state $|R^{\uparrow}\rangle$ or $|L^{\downarrow}\rangle$ (the upper arrow indicates the direction in which the photon travels along the Z-axis), but if the incoming photon is in state $|R^{\downarrow}\rangle$ or $|L^{\uparrow}\rangle$, the photon enters the cavity and is directly transmitted out, with only one phase change. The details are described as follows:

If three photons, independent of each other, are in the same quantum state $|\Phi^{ph}\rangle_{a,b,c} = (|R\rangle_{a,b,c} + |L\rangle_{a,b,c})/\sqrt{2}$, they each enter the cavity field in turn, when the atoms in the cavity are in state $|\Phi^{s}\rangle = (|\uparrow\rangle + |\downarrow\rangle)/\sqrt{2}$. The incoming photons interact with the atoms in the cavity and become entangled. Four scenarios are analyzed.

1. If three photons are emitted into the cavity field and interact with atoms in the cavity field, all three photons are projected out of the cavity field, then the entangled state generated after the interaction between the incoming photons and atoms in the cavity is

$$|\Phi^{1}\rangle \to -t^{3}(w)/2\sqrt{2}(|R\rangle_{a}|R\rangle_{b}|R\rangle_{c}|\downarrow\rangle + |L\rangle_{a}|L\rangle_{b}|L\rangle_{c}|\uparrow\rangle)/\sqrt{2}. \quad (4.67)$$

2. If three photons are emitted into the cavity field and interact with atoms in the cavity field, all three photons are reflected, then the entangled state generated after the interaction between the incoming photons and atoms in the cavity is

$$|\Phi^{2}\rangle \to r^{3}(w)/2\sqrt{2}(|R\rangle_{a}|R\rangle_{b}|R\rangle_{c}|\downarrow\rangle + |L\rangle_{a}|L\rangle_{b}|L\rangle_{c}|\uparrow\rangle)/\sqrt{2}. \quad (4.68)$$

3. If three photons are emitted into the cavity field and interact with atoms in the cavity field, one of them is reflected and the other two are projected out of the cavity, then the entangled state generated after the interaction between the incoming photon and atoms in the cavity is

$$|\Phi^{3}\rangle \to 3t^{2}(w)r(w)/2\sqrt{2}(|R\rangle_{a}|R\rangle_{b}|R\rangle_{c}|\downarrow\rangle + |L\rangle_{a}|L\rangle_{b}|L\rangle_{c}|\uparrow\rangle)/\sqrt{2}. \quad (4.69)$$

4. If three photons are emitted into the cavity field and interact with atoms in the cavity field, two of them are reflected and one photon is projected out of the cavity, then the entangled state generated after the interaction between the incoming photon and atoms in the cavity is

$$|\Phi^{4}\rangle \to -3t(w)r^{2}(w)/2\sqrt{2}(|R\rangle_{a}|R\rangle_{b}|R\rangle_{c}|\downarrow\rangle + |L\rangle_{a}|L\rangle_{b}|L\rangle_{c}|\uparrow\rangle)/\sqrt{2}. \quad (4.70)$$

Thus, regardless of whether the photon is projected or reflected, the resulting entangled state can be regarded as

$$|\Phi\rangle \rightarrow (|R\rangle_a|R\rangle_b|R\rangle_c|\downarrow\rangle + |L\rangle_a|L\rangle_b|L\rangle_c|\uparrow\rangle)/\sqrt{2}. \tag{4.71}$$

Then the atom in the cavity is measured after the H-gate operation is performed on the atom. If the atom is in state $|\uparrow\rangle$, the entangled state we get is

$$|\Phi^+\rangle = (|R\rangle_a|R\rangle_b|R\rangle_c + |L\rangle_a|L\rangle_b|L\rangle_c)/\sqrt{2}. \tag{4.72}$$

If the atom is in state $|\downarrow\rangle$, the entangled state we get is

$$|\Phi^-\rangle = (|R\rangle_a|R\rangle_b|R\rangle_c - |L\rangle_a|L\rangle_b|L\rangle_c)/\sqrt{2}. \tag{4.73}$$

Combining Formulas (4.72) and (4.73), we get the GHZ state as follows:

$$|\Phi^\pm\rangle = (|R\rangle_a|R\rangle_b|R\rangle_c \pm |L\rangle_a|L\rangle_b|L\rangle_c)/\sqrt{2}. \tag{4.74}$$

4.4 Summary of This Chapter

In this chapter, two kinds of quantum information splitting schemes are proposed, one of which uses a four-qubit entangled state as the quantum channel to split the two-qubit Bell state and performs the measurement of the GHZ state. Finally, the information receiver successfully receives the sent information without information leakage. Under the assumption of external attack and internal attack, the analysis results show that neither external attack nor internal attack has any chance to obtain the information sent by Alice. Another kind of quantum information splitting scheme is using the five-qubit entangled state to split arbitrary one-qubit and two-qubit state, using a four-qubit cluster state and a two-qubit Bell state to split arbitrary two-qubit state, using a four-qubit cluster state and GHZ state to split any three-qubit state. In this kind of quantum information splitting process, Alice, the information sender, performs Bell state measurement of different qubit pairs, Charlie performs single-qubit measurement under different state bases, and finally, Bob, the information receiver, performs the appropriate unitary operation to reconstruct the original state information. This kind of information splitting process is also physically realized in QED, which provides a good communication basis for the quantum teleportation protocol that is immune to noise in the later stage.

Chapter 5
Controllable Multiple Degree of Freedom Quantum Teleportation Protocol for Immune Noise

Combined with the entanglement source preparation and Bell state measurement characteristics in noise channel, there have been many studies on the theory and experiment of quantum teleportation. However, the classical quantum teleportation is based on the independent assumption of a single degree of freedom, which cannot meet the requirements of long-distance robust quantum teleportation. Under the condition of long-distance quantum communication network, quantum entanglement and measurement have diversity and variability, and the overall demand for quantum physical system is not the accumulation of a single degree of freedom. Quantum teleportation has the characteristics of long-distance transmission and density. In the real quantum physical system, quantum has multiple degrees of freedom. Moreover, quantum resources not only are pure states but also include entangled states, mixed states, coherent states, etc. The teleportation performance of a quantum depends not only on the preparation of the entangled source but also on a series of measurements.

Therefore, in Chap. 4, we have completed the splitting of Bell state and arbitrary state information in different channels and obtained the quantum resources required for quantum teleportation. In this chapter, we introduce the new characteristics of spin-orbit angular momentum and super Bell state measurement resources into the quantum teleportation theory. By analyzing the distribution of the multi-degree of freedom demand intensity in time, space, and frequency, we design the multi-degree of freedom quantum body teleportation protocol and achieve the goal of meeting the multidemand of super Bell state measurement under the constraints of their respective angular momentum. Figure 5.1 shows the framework of the multi-degree of freedom quantum system teleportation protocol scheme.

In the process of protocol design, we combine the research methods of quantum information splitting and measurement and quantum coding with the classical basic theory of quantum teleportation. Based on the measurement and analysis of quantum information splitting, we found the conversion relationship between spin-orbit angular momentum and hyperentangled Bell state. Based on the graph state concatenated coding analysis method of quantum coding, we can quickly calculate

D. Li, *Quantum Teleportation Under Noise*,
https://doi.org/10.1007/978-3-031-91250-4_5

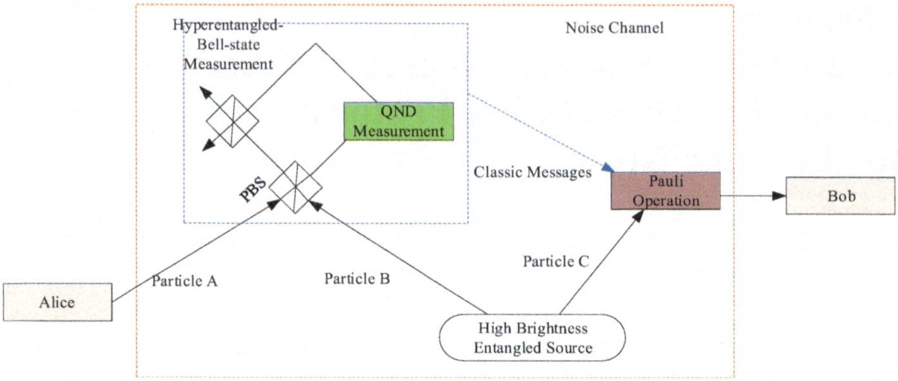

Fig. 5.1 Scheme framework of multi-degree of freedom quantum teleportation protocol

the coherent information, the approximation value of channel capacity, and the noise tolerance of various concatenated codes in the channel, and obtain the regions where quantum information can be transmitted in different noise channels. Furthermore, we find the progressive relationship between the preparation of high-brightness entangled source and the measurement of Bell state basis through the characteristics of multiple quantum beams. On this basis, we combine matrix decomposition and other analysis methods with local measurement and dense measurement methods to improve the efficiency and security of measurement.

5.1 Preparation of High-Brightness Entangled Source

The key of the multi-degree of freedom quantum system teleportation protocol is to solve the problems of preparation, coding, and Bell state measurement of multi-degree of freedom high-brightness entangled source under spin-orbit angular momentum. So how to prepare a high-brightness entangled source suitable for multi-degree of freedom (orbit spin angular momentum), and measure the hyperentangled Bell state; distinguish 1 of the 16 Bell states from the other 15; carry out basic measurement, enhanced local measurement, and ultra-dense measurement of the hyperentangled Bell state; and carry out controllable identity authentication of quantum teleportation, so as to achieve the purpose of quantum teleportation in multi-degree of freedom systems, is a key problem.

At present, most quantum teleportation protocols are only based on the quantum state of a single degree of freedom. However, we have found that the real quantum physical system has multiple degrees of freedom, such as the simplest single photon, whose properties mainly include wavelength, momentum, spin angular momentum, and orbital angular momentum. On the basis of the proposed immune noise model and information splitting, considering that quantum teleportation is a linear operation of quantum states, we study the multi-degree of freedom characteristics of quantum physical systems. Through the characteristics of single-photon teleportation under spin angular momentum and orbital angular momentum, we design the preparation scheme of high-brightness entangled source under multi-degree of freedom.

Specifically, for example, Alice wants to transmit the mixed quantum state a information of a single particle to Bob. We encode the spin angular momentum (*SAM*) and orbital angular momentum (*SAM*):

$$|\varphi\rangle_a = \alpha|0\rangle_a^S|0\rangle_a^o + \beta|0\rangle_a^S|1\rangle_a^o + \gamma|1\rangle_a^S|0\rangle_a^o + \delta|1\rangle_a^S|1\rangle_{a,}^o \tag{5.1}$$

satisfying $|\alpha|^2 + |\beta|^2 + |\gamma|^2 + |\delta|^2 = 1$. And Alice and Bob must share a pair of hyperentangled particle pairs (b, c), which must be entangled in both spin angular momentum and orbital angular momentum, that is,

$$|\partial\rangle_{bc} = |\Phi^-\rangle_{bc}|\varpi^+\rangle_{bc} = \frac{1}{2}\left(|0\rangle_b^S|0\rangle_c^S - |1\rangle_b^S|1\rangle_c^S\right)\left(|0\rangle_b^o|0\rangle_c^o + |1\rangle_b^o|1\rangle_c^o\right). \tag{5.2}$$

Among them, the four Bell states encoded under spin angular momentum are represented by $|\Phi^\pm\rangle = \frac{1}{\sqrt{2}}(|0\rangle^S|0\rangle^S \pm |1\rangle^S|1\rangle^S)$ and $|\Psi^\pm\rangle = \frac{1}{\sqrt{2}}(|0\rangle^S|1\rangle^S \pm |1\rangle^S|0\rangle^S)$. The four Bell states under orbital angular momentum are represented by $|\varpi^\pm\rangle = \frac{1}{\sqrt{2}}(|0\rangle^o|0\rangle^o \pm |1\rangle^o|1\rangle^o)$ and $|\omega^\pm\rangle = \frac{1}{\sqrt{2}}(|0\rangle^o|1\rangle^o \pm |1\rangle^o|0\rangle^o)$.

The orthogonal hyperentangled Bell states of particles a, b, c are described as follows:

$$\begin{aligned}
&|\Theta\rangle^1 = \frac{1}{4}(|\Phi^+\rangle|\varpi^+\rangle)_{ab}\sigma_{cz}^s|\varphi\rangle_c, \quad |\Theta\rangle^2 = \frac{1}{4}(|\Phi^+\rangle|\varpi^-\rangle)_{ab}\sigma_{cz}^s\sigma_{cz}^o|\varphi\rangle_c,\\
&|\Theta\rangle^3 = \frac{1}{4}(|\Phi^+\rangle|\omega^+\rangle)_{ab}\sigma_{cz}^s\sigma_{cx}^o|\varphi\rangle_c, \quad |\Theta\rangle^4 = -\frac{1}{4}(|\Phi^+\rangle|\omega^-\rangle)_{ab}i\sigma_{cz}^s\sigma_{cy}^o|\varphi\rangle_c,\\
&|\Theta\rangle^5 = \frac{1}{4}(|\Phi^-\rangle|\varpi^+\rangle)_{ab}|\varphi\rangle_c, \quad |\Theta\rangle^6 = \frac{1}{4}(|\Phi^-\rangle|\varpi^-\rangle)_{ab}\sigma_{cz}^o|\varphi\rangle_c,\\
&|\Theta\rangle^7 = \frac{1}{4}(|\Phi^-\rangle|\omega^+\rangle)_{ab}\sigma_{cx}^o|\varphi\rangle_c, \quad |\Theta\rangle^8 = -\frac{1}{4}(|\Phi^-\rangle|\omega^-\rangle)_{ab}i\sigma_{cy}^o|\varphi\rangle_c,\\
&|\Theta\rangle^9 = \frac{1}{4}(|\Psi^+\rangle|\varpi^+\rangle)_{ab}i\sigma_{cy}^s|\varphi\rangle_c, \quad |\Theta\rangle^{10} = \frac{1}{4}(|\Psi^+\rangle|\varpi^-\rangle)_{ab}i\sigma_{cy}^s\sigma_{cz}^o|\varphi\rangle_c,\\
&|\Theta\rangle^{11} = \frac{1}{4}(|\Psi^+\rangle|\omega^+\rangle)_{ab}i\sigma_{cy}^s\sigma_{cx}^o|\varphi\rangle_c, \quad |\Theta\rangle^{12} = \frac{1}{4}(|\Psi^+\rangle|\omega^-\rangle)_{ab}\sigma_{cy}^s\sigma_{cx}^o|\varphi\rangle_c,\\
&|\Theta\rangle^{13} = -\frac{1}{4}(|\Psi^-\rangle|\varpi^+\rangle)_{ab}\sigma_{cx}^s|\varphi\rangle_c, \quad |\Theta\rangle^{14} = -\frac{1}{4}(|\Psi^-\rangle|\varpi^-\rangle)_{ab}\sigma_{cx}^s\sigma_{cz}^o|\varphi\rangle_c,\\
&|\Theta\rangle^{15} = -\frac{1}{4}(|\Psi^-\rangle|\omega^+\rangle)_{ab}\sigma_{cx}^s\sigma_{cx}^o|\varphi\rangle_c, \quad |\Theta\rangle^{16} = \frac{1}{4}(|\Psi^-\rangle|\omega^-\rangle)_{ab}i\sigma_{cx}^s\sigma_{cy}^o|\varphi\rangle_c.
\end{aligned} \tag{5.3}$$

This shows that for the unknown state $|\varphi\rangle_a$, we have obtained 16 possible measurement results. By measuring the hyperentangled Bell state of particle a and particle b, Alice can project the particle r onto these 16 collapsed states. Then, Alice tells Bob the measurement results of the hyperentangled Bell state through the classical channel, and Bob converts the particles into the original state information by performing the appropriate two-particle local unitary operation.

5.2 Measurement of Hyperentangled Bell States and Ultra-dense Coding

In Sect. 5.1, after the violet femtosecond pulse is focused on three BBO crystals to obtain a multi-degree of freedom high-brightness entanglement source, in this section, we measure and code the hyperentangled ion state under multi-degrees of freedom. For example, a particle state can be entangled in two or more degrees of freedom, such as polarization angular momentum, polarization energy, and polarization time-space mode. Due to the large information carried by hyperentanglement, it can entangle multiple degrees of freedom and occupy less resources, which effectively improves the channel capacity and facilitates long-distance control. Moreover, the measurement and control do not affect or destroy the particle states in other degrees of freedom. For example, any two degrees of freedom of a photon can be entangled with the two degrees of freedom of another photon at the same time, so a photon can be used as two qubits.

The hyperentangled state in the polarization space degree of freedom is expressed as follows:

$$\left|\Phi_{AB}^{+}\right\rangle_{PS} = \frac{1}{2}(|00\rangle + |11\rangle)_P + (|a_1b_1\rangle + |a_2b_2\rangle)_S, \tag{5.4}$$

where the subscript P represents the polarization degree of freedom and the subscript S represents the spatial mode degree of freedom. $|0\rangle$ represents the horizontal polarization state of photons in the degree of polarization freedom, and $|1\rangle$ represents the vertical polarization state of photons in the degree of polarization freedom.

Because the difficulty of preparing multiphoton entangled states increases exponentially with the number of photons, the brightness and fidelity are very high. And it is very easy to couple the polarization mode and spatial mode of two photons in EPR entangled pair to prepare a two-dimensional 4-bit hyperentangled state.

The two-dimensional 4-bit hyperentangled state in the polarization space degree of freedom is expressed as follows:

$$\left|\Phi_{AB}^{+}\right\rangle_{PS} = \frac{1}{2}(|0000\rangle + |0011\rangle + |1100\rangle + |1111\rangle)_P + (|a_1b_1\rangle + |a_2b_2\rangle)_S, \tag{5.5}$$

where the subscript P represents the polarization degree of freedom and the subscript S represents the spatial mode degree of freedom. $|0\rangle$ represents the horizontal polarization state of photons in the degree of polarization freedom, and $|1\rangle$ represents the vertical polarization state of photons in the degree of polarization freedom.

For a two-dimensional four-qubit hyperentangled state quantum system with polarization space degrees of freedom, there are 16 Bell states, which are expressed as follows:

$$|\Omega_{AB}\rangle_{PS} = |\mathfrak{R}\rangle_P \otimes |\aleph\rangle_S,$$

where $|\mathfrak{R}\rangle_P$ represents four Bell states under the degree of freedom of polarization (A and B represent two photons, respectively):

$$\begin{aligned} |\Phi^+\rangle_P &= \frac{1}{\sqrt{2}}(|00\rangle+|11\rangle)_{AB} \quad |\Phi^-\rangle_P = \frac{1}{\sqrt{2}}(|00\rangle-|11\rangle)_{AB,} \\ |\Psi^+\rangle_P &= \frac{1}{\sqrt{2}}(|01\rangle+|10\rangle)_{AB} \quad |\Psi^-\rangle_P = \frac{1}{\sqrt{2}}(|01\rangle-|10\rangle)_{AB.} \end{aligned} \tag{5.6}$$

$|\aleph\rangle_S$ represents the four Bell states under spatial degrees of freedom (A and B represent two photons, respectively):

$$\begin{aligned} |\Phi^+\rangle_S &= \frac{1}{\sqrt{2}}(|a_1b_1\rangle+|a_2b_2\rangle)_{AB} \quad |\Phi^-\rangle_S = \frac{1}{\sqrt{2}}(|a_1b_1\rangle-|a_2b_2\rangle)_{AB}, \\ |\Psi^+\rangle_S &= \frac{1}{\sqrt{2}}(|a_1b_2\rangle+|a_2b_1\rangle)_{AB} \quad |\Psi^-\rangle_S = \frac{1}{\sqrt{2}}(|a_1b_2\rangle-|a_2b_1\rangle)_{AB.} \end{aligned} \tag{5.7}$$

Therefore, $|\mathfrak{R}\rangle_P$ and $|\aleph\rangle_S$ are immune to combined noise.

The two non-orthogonal measurement bases under polarization degrees of freedom are $\{|0\rangle, |1\rangle\}$ and $\left\{|\pm\rangle_P = \frac{1}{\sqrt{2}}(|0\rangle \pm |1\rangle)\right\}$. The two non-orthogonal measurement bases under spatial degrees of freedom are $\{|a_1\rangle, |a_2\rangle\}$ and $\left\{|\rangle_S = \frac{1}{\sqrt{2}}(|a_1\rangle \pm |a_2\rangle)\right\}$.

5.2.1 Hyper-Bell State Measurement Method

We first distinguish one of the 16 spin-orbit angular momentum hyperentangled Bell states from the other 15 by using the basic measurement technique of super-entangled Bell states; Second, an enhanced local measurement method is designed to improve the efficiency of Bell state measurement; Third, the ultra-dense state measurement method is designed to improve the security of quantum teleportation by using the third-party entanglement measurement. As shown in Fig. 5.2, the research scheme of the hyper-Bell state measurement method is shown.

Based on the research on the preparation of high-brightness entangled source with multiple degrees of freedom, we describe the progressive relationship between entangled source and measurement and design the basic measurement, enhanced local measurement, and ultra-dense measurement models of hyperentangled Bell

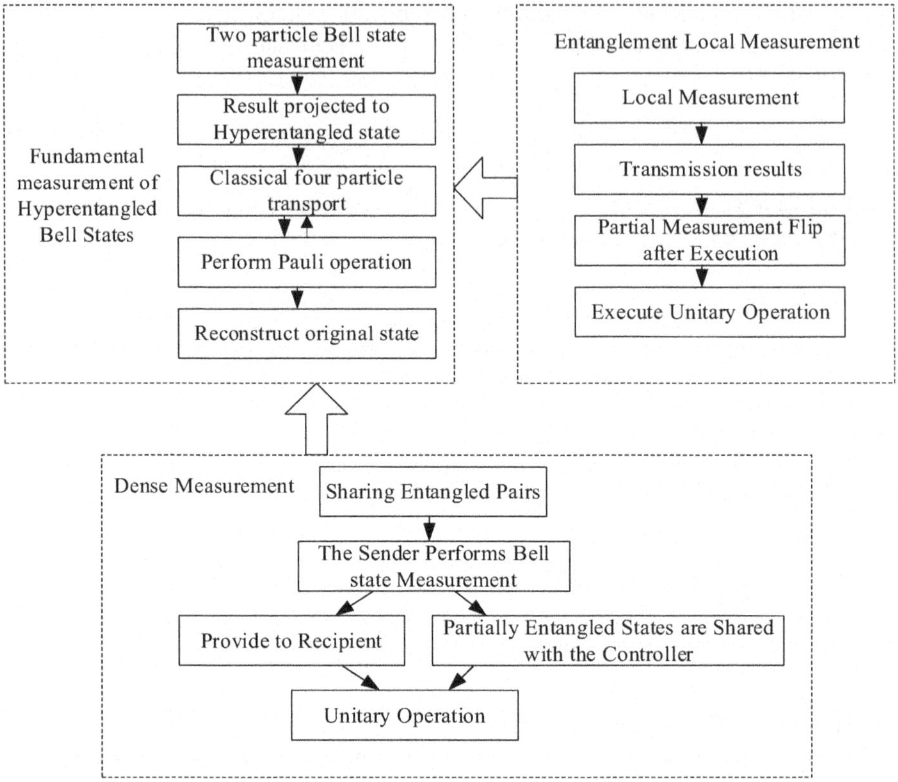

Fig. 5.2 Hyperentangled Bell state measurement method

states, which lays a good foundation for the realization of teleportation of secure quantum systems in multiple degrees of freedom.

5.2.1.1 Basic Measurement

In order to carry out quantum teleportation under multiple degrees of freedom, Bell state measurement must be carried out. Firstly, we designed the basic measurement model of the hyperentangled Bell state and distinguished one of the spin angular momentum and orbital angular momentum hyperentangled Bell states from the other by using the characteristics of the polarization beam splitter.

Specifically, Alice first measured the two particle Bell states of particle *A* and particle *B* and projected the measurement results onto 16 groups of super-entangled Bell states, one of which is

$$\left(|\vartheta\rangle_{AB} = |\Phi^-\rangle_{AB}|\phi^+\rangle_{AB} = \frac{1}{2}\left(|0\rangle_A^s|0\rangle_B^s - |1\rangle_{\mathrm{A}}^{\mathrm{s}}|1\rangle_B^s\right)\left(0\rangle_A^0|0\rangle_B^0 + |1\rangle_A^0|1\rangle_B^0\right). \tag{5.8}$$

This process is regarded as the measurement of hyperentangled Bell states (h-bsm). After the measurement of the hyperentangled Bell state, particle C will be projected onto the original state particle A, i.e.,

$$|\varphi\rangle_C = \alpha|0\rangle_C^s|0\rangle_C^o + \beta|0\rangle_C^s|1\rangle_C^o + \gamma|1\rangle_C^s|0\rangle_C^o + \delta|1\rangle_{C^s}|1\rangle_C^o. \tag{5.9}$$

There are 16 possible outcomes in this case. Of course, the particles (A, B) will also be projected onto the other 15 Bell states, and the results will be transmitted through the classical four particles. Bob performs an appropriate Pauli operation to reconstruct the mixed state information of particle A.

5.2.1.2 Enhanced Local Measurement

On the basis of the basic measurement of hyperentangled Bell states, in order to improve the measurement efficiency, we designed an enhanced local measurement strategy. Its architecture is shown in Fig. 5.3.

Specifically, we let particle B go through three stages: preferential particle measurement, process noise, and post-part flip measurement. Before Alice sends a message to Bob, she first performs a priority local measurement (intensity P) on the two qubits. After Bob receives the measurement results, he first performs the subsequent partial measurement reversal (i.e., bit reversal and local measurement) and then performs the local unitary operation in turn. Finally, the entangled state is obtained:

$$\rho_B = M_0^{-1}[\wedge(M_0\rho_0 M_0\dagger)](M_0^{-1}), \text{where } \ M_0{}^{-1} = I_1 \otimes m_0{}^{-1}. \tag{5.10}$$

Enhancing the intensity ρ of particle measurement, the reversed state $|00\rangle$ is closer to the original EPR state, and it is immune to noise, thus improving the measurement efficiency and safety.

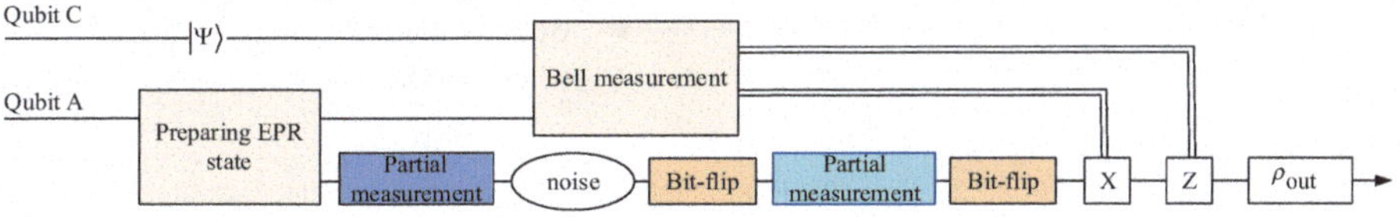

Fig. 5.3 Enhanced local measurement

5.2.1.3 Ultra-dense Measurement

In order to strengthen the measurement in multi-degrees of freedom, we propose an ultra-dense measurement scheme that controls the information of the sender and the receiver, so we create the entangled state. Our specific scheme is shown in Fig. 5.4.

Alice, the information sender, and Bob, the information receiver, share an entangled pair in advance and use some technologies to transmit quantum state information on the classical information channel. In the single-particle quantum teleportation, Alice performs Bell state measurement (i.e., the maximum entangled state) based on the unknown state and provides this state to Charlie and shares some of his entangled states with Bob. She sends the measurement results of two particles to Bob on the classical channel. On the basis of Alice information, Bob performs one of the four unitary operations on his half-entangled pairs and transmits the operation results to Charlie.

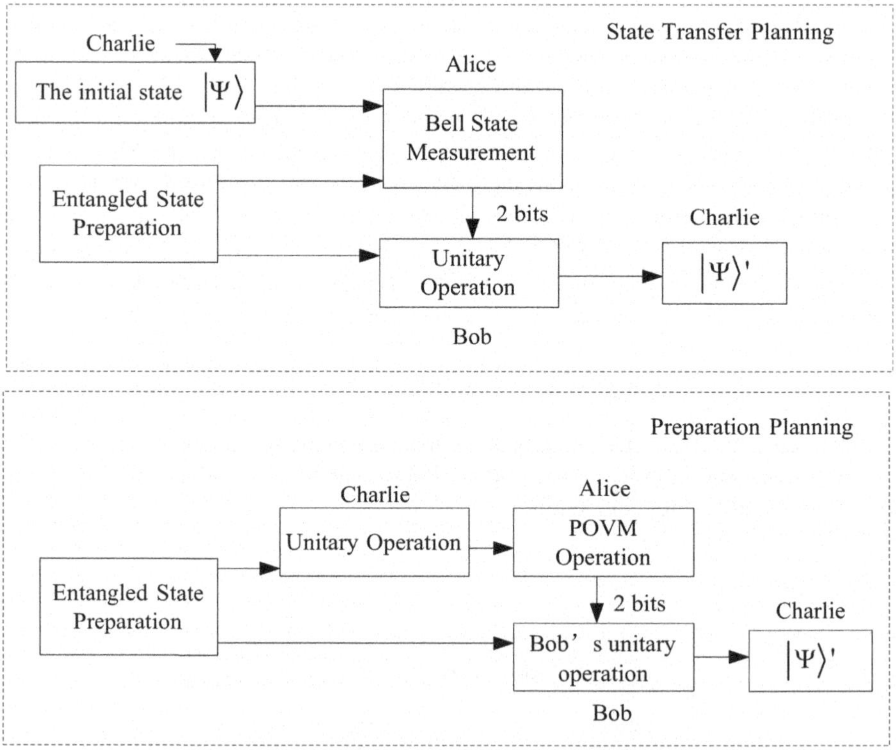

Fig. 5.4 Ultra-dense measurement strategy

5.2.2 *Bell State Ultra-dense Coding*

Due to the insufficient information of each quantum bit in the process of quantum teleportation coding, this will affect the channel capacity and efficiency of quantum teleportation. Most traditional coding methods are based on entangled states and cannot distinguish between Bell states and GHZ states. The channel capacity does not reach the theoretical value and cannot meet the needs of long-distance quantum teleportation. In order to solve this problem, we use the two-dimensional four-qubit hyperentangled state as the quantum carrier, and we need to realize the dense coding of the two-dimensional four-qubit hyperentangled state under the joint dephasing noise and the joint rotation noise, respectively.

5.2.2.1 Super-dense Coding Under Dephase Noise

When suffering from collective dephasing noise, the horizontal photon $|0\rangle$ remains unchanged and the vertical photon $|1\rangle$ becomes $e^{i\varphi} \mid 1\rangle$. The two logical qubits can resist the noise. We define it as follows:

$$|0\rangle_{dp} = |01\rangle, \quad |1\rangle_{dp} = |10\rangle. \tag{5.11}$$

Their superposition states can also resist noise, which is defined as follows:

$$\begin{aligned} |+\rangle_{dp} &= \frac{1}{\sqrt{2}}\left(|0\rangle_{dp} + |1\rangle_{dp}\right) = \frac{1}{\sqrt{2}}(|01\rangle + |10\rangle), \\ |-\rangle_{dp} &= \frac{1}{\sqrt{2}}\left(|0\rangle_{dp} - |1\rangle_{dp}\right) = \frac{1}{\sqrt{2}}(|01\rangle - |10\rangle). \end{aligned} \tag{5.12}$$

The four original Bell states are described as follows:

$$\begin{aligned} |\Phi^+\rangle &= \frac{1}{\sqrt{2}}(|00\rangle + |11\rangle), |\Phi^-\rangle = \frac{1}{\sqrt{2}}(|00\rangle - |11\rangle), \\ |\Psi^+\rangle &= \frac{1}{\sqrt{2}}(|01\rangle + |10\rangle), |\Psi^-\rangle = \frac{1}{\sqrt{2}}(|01\rangle - |10\rangle). \end{aligned} \tag{5.13}$$

From the above formula, we can see that the four logical Bell states that resist noise are written as follows:

$$\begin{aligned}\left|\Phi_{dp}^{+}\right\rangle_{1234} &= \frac{1}{\sqrt{2}}\left(\left|0_{dp}\right\rangle\left|0_{dp}\right\rangle + \left|1_{dp}\right\rangle\left|1_{dp}\right\rangle\right)_{1234} = \frac{1}{\sqrt{2}}(|01\rangle|01\rangle + |10\rangle|10\rangle)_{1234} \\ &= \frac{1}{\sqrt{2}}(|00\rangle|11\rangle + |11\rangle|00\rangle)_{1324} = \frac{1}{\sqrt{2}}(|\Phi^{+}\rangle|\Phi^{+}\rangle - |\Phi^{-}\rangle|\Phi^{-}\rangle)_{1324},\end{aligned} \tag{5.14}$$

$$\begin{aligned}\left|\Phi_{dp}^{-}\right\rangle_{1234} &= \frac{1}{\sqrt{2}}\left(\left|0_{dp}\right\rangle\left|0_{dp}\right\rangle - \left|1_{dp}\right\rangle\left|1_{dp}\right\rangle\right)_{1234} = \frac{1}{\sqrt{2}}(|01\rangle|01\rangle - |10\rangle|10\rangle)_{1234} \\ &= \frac{1}{\sqrt{2}}(|00\rangle|11\rangle - |11\rangle|00\rangle)_{1324} = \frac{1}{\sqrt{2}}(|\Phi^{-}\rangle|\Phi^{+}\rangle - |\Phi^{+}\rangle|\Phi^{-}\rangle)_{1324},\end{aligned} \tag{5.15}$$

$$\begin{aligned}\left|\Psi_{dp}^{+}\right\rangle_{1234} &= \frac{1}{\sqrt{2}}\left(\left|0_{dp}\right\rangle\left|1_{dp}\right\rangle + \left|1_{dp}\right\rangle\left|0_{dp}\right\rangle\right)_{1234} = \frac{1}{\sqrt{2}}(|01\rangle|10\rangle + |10\rangle|01\rangle)_{1234} \\ &= \frac{1}{\sqrt{2}}(|01\rangle|10\rangle + |10\rangle|01\rangle)_{1324} = \frac{1}{\sqrt{2}}(|\Psi^{+}\rangle|\Psi^{+}\rangle - |\Psi^{-}\rangle|\Psi^{-}\rangle)_{1324},\end{aligned} \tag{5.16a}$$

$$\begin{aligned}\left|\Psi_{dp}^{-}\right\rangle_{1234} &= \frac{1}{\sqrt{2}}\left(\left|0_{dp}\right\rangle\left|1_{dp}\right\rangle - \left|1_{dp}\right\rangle\left|0_{dp}\right\rangle\right)_{1234} = \frac{1}{\sqrt{2}}(|01\rangle|10\rangle - |10\rangle|01\rangle)_{1234} \\ &= \frac{1}{\sqrt{2}}(|01\rangle|10\rangle - |10\rangle|01\rangle)_{1324} = \frac{1}{\sqrt{2}}(|\Psi^{-}\rangle|\Psi^{+}\rangle - |\Psi^{+}\rangle|\Psi^{-}\rangle)_{1324}.\end{aligned} \tag{5.17}$$

The four original unitary operations are described as follows:

$$\begin{aligned}&I = |00\rangle + |11\rangle = |0\rangle\langle 0| + |1\rangle\langle 1|, U_x = |10\rangle + |01\rangle = |1\rangle\langle 0| + |0\rangle\langle 1|, \\ &U_u = |01\rangle - |10\rangle = |0\rangle\langle 1| - |1\rangle\langle 0|, U_z = |00\rangle - |11\rangle = |0\rangle\langle 0| - |1\rangle\langle 1|.\end{aligned} \tag{5.16b}$$

From the original unitary operation, we get four logical unitary operations immune to noise, as follows:

$$\begin{aligned}&\Pi_{00} = \Pi_I = I_1 \otimes I_2, \quad \Pi_{01} = \Pi_z = U_{z1} \otimes I_2, \\ &\Pi_{10} = \Pi_x = U_{x1} \otimes U_{x2}, \quad \Pi_{11} = \Pi_y = U_{y1} \otimes U_{x2}.\end{aligned} \tag{5.18}$$

The logical Bell state ultra-dense coding is shown in Table 5.1.

5.2.2.2 Ultra-dense Coding in Rotating Noise Channel

In the collective rotation noise channel, the horizontal photon $|0\rangle$ becomes $\cos\theta \mid 0\rangle + \sin\theta \mid 1\rangle$, the vertical photon $|1\rangle$ becomes $-\sin\theta \mid 0\rangle + \cos\theta \mid 1\rangle$, and the two logical qubits are immune to rotation noise. We define it as follows:

Table 5.1 Logical Bell state ultra-dense coding

	$\left\|\Phi_{d_n}^+\right\rangle$	$\left\|\Phi_{do}^-\right\rangle$	$\left\|\Psi_{d_n}^+\right\rangle$	$\left\|\Psi_{do}^-\right\rangle$
Π_I	$\left\|\Phi_{d_p}^+\right\rangle$	$\left\|\Phi_{d_p}^-\right\rangle$	$\left\|\Psi_{d_p}^+\right\rangle$	$\left\|\Psi_{d_p}^-\right\rangle$
Π_Z	$\left\|\Phi_{d_p}^-\right\rangle$	$\left\|\Phi_{d_p}^+\right\rangle$	$\left\|\Psi_{d_p}^-\right\rangle$	$\left\|\Psi_{d_n}^+\right\rangle$
Π_X	$\left\|\Psi_{d_p}^+\right\rangle$	$\left\|\Psi_{d_n}^-\right\rangle$	$\left\|\Phi_{d_p}^+\right\rangle$	$\left\|\Phi_{do}^-\right\rangle$
Π_V	$\left\|\Psi_{d_p}^-\right\rangle$	$\left\|\Psi_{d_p}^+\right\rangle$	$\left\|\Phi_{d_p}^-\right\rangle$	$\left\|\Phi_{d_p}^+\right\rangle$

$$|0\rangle_{dp} = |\Phi^+\rangle, \quad |1\rangle_{dp} = |\psi^-\rangle. \tag{5.19}$$

Their superposition states are also immune to joint rotation noise, which we define as follows:

$$\begin{aligned} |+\rangle_r &= \frac{1}{\sqrt{2}}(|0\rangle_r + |1\rangle_r) = \frac{1}{\sqrt{2}}(|\Phi^+\rangle + |\psi^-\rangle), \\ |-\rangle_r &= \frac{1}{\sqrt{2}}(|0\rangle_r - |1\rangle_r) = \frac{1}{\sqrt{2}}(|\Phi^+\rangle - |\psi^-\rangle). \end{aligned} \tag{5.20}$$

These four logical Bell states are immune to joint rotation noise, which we write as follows:

$$\begin{aligned} |\Phi_r^+\rangle_{1234} &= \frac{1}{\sqrt{2}}(|0_r\rangle|0_r\rangle + |1_r\rangle|1_r\rangle)_{1234} = \frac{1}{\sqrt{2}}(|\Phi^+\rangle|\Phi^+\rangle + |\Psi^-\rangle|\Psi^-\rangle_{1234} \\ &= \frac{1}{\sqrt{2}}(|\Phi^+\rangle|\Phi^+\rangle + |\Psi^-\rangle|\Psi^-\rangle_{1324}, \end{aligned} \tag{5.21}$$

$$\begin{aligned} |\Phi_r^-\rangle_{1234} &= \frac{1}{\sqrt{2}}(|0_r\rangle|0_r\rangle - |1_r\rangle|1_r\rangle)_{1234} = \frac{1}{\sqrt{2}}(|\Phi^+\rangle|\Phi^+\rangle|\Psi^-\rangle|\Psi^-\rangle_{1234} \\ &= \frac{1}{\sqrt{2}}(|\Phi^-\rangle|\Phi^-\rangle + |\Psi^+\rangle|\Psi^+\rangle_{1324}, \end{aligned} \tag{5.22}$$

$$\begin{aligned} |\Psi_r^+\rangle_{1234} &= \frac{1}{\sqrt{2}}(|0_r\rangle|1_r\rangle + |1_r\rangle|0_r\rangle)_{1234} = \frac{1}{\sqrt{2}}(|\Phi^+\rangle|\Psi^-\rangle + |\Psi^-\rangle|\Phi^+\rangle_{1234} \\ &= \frac{1}{\sqrt{2}}(|\Phi^-\rangle|\Psi^+\rangle - |\Psi^+\rangle|\Phi^-\rangle_{1324}, \end{aligned} \tag{5.23}$$

Table 5.2 Logical Bell state ultra-dense coding

	$\vert\Phi_r^+\rangle$	$\vert\Phi_r^-\rangle$	$\vert\Psi_r^+\rangle$	$\vert\Psi_r^-\rangle$
Ξ_I	$\vert\Phi_r^+\rangle$	$\vert\Phi_r^-\rangle$	$\vert\Psi_r^+\rangle$	$\vert\Psi_r^-\rangle$
Ξ_T	$\vert\Phi_r^-\rangle$	$\vert\Phi_r^+\rangle$	$\vert\Psi_r^-\rangle$	$\vert\Psi_r^+\rangle$
Ξ_X	$\vert\Psi_r^+\rangle$	$\vert\Psi_r^-\rangle$	$\vert\Phi_r^+\rangle$	$\vert\Phi_r^-\rangle$
Ξ_Y	$\vert\Psi_r^-\rangle$	$\vert\Psi_r^+\rangle$	$\vert\Phi_r^-\rangle$	$\vert\Phi_r^+\rangle$

$$
\begin{aligned}
|\Psi_r^+\rangle_{1234} &= \frac{1}{\sqrt{2}}(|0_r\rangle|1_r\rangle - |1_r\rangle|0_r\rangle)_{1234} = \frac{1}{\sqrt{2}}(|\Phi^+\rangle|\Psi^-\rangle - |\Psi^-\rangle|\Phi^+\rangle_{1234} \\
&= \frac{1}{\sqrt{2}}(|\Phi^+\rangle|\Psi^-\rangle - |\Psi^-\rangle|\Phi^+\rangle_{1324}.
\end{aligned} \tag{5.24}
$$

The logic unitary operation of four immune noises is obtained as follows:

$$
\begin{aligned}
&\Xi_{00} = \Xi_I = I_1 \otimes I_2, \quad \Xi_{01} = \Xi_z = U_{z1} \otimes I_2, \\
&\Xi_{10} = \Xi_x = U_{z1} \otimes U_{x2}, \quad \Xi_{11} = \Xi_y = I_1 \otimes U_{y2}.
\end{aligned} \tag{5.25}
$$

Therefore, the logic Bell state super-dense coding in this case is shown in Table 5.2.

5.2.2.3 Ultra-dense Coding Under Joint Common Noise

In the joint dephasing (rotation) noise channel, we describe the two original logical Bell states of entanglement swapping as follows:

$$
\begin{aligned}
\left|\Phi_{dp/r}^+\right\rangle_{x_1y_1} \otimes \left|\Phi_{dp/r}^+\right\rangle_{x_2y_2} &= \frac{1}{2}\Big(\left|\Phi_{dp/r}^+\right\rangle_{x_1x_2} \otimes \left|\Phi_{dp/r}^+\right\rangle_{y_1y_2} + \left|\Phi_{dp/r}^-\right\rangle_{x_1x_2} \otimes \left|\Phi_{dp/r}^-\right\rangle_{y_1y_2} \\
&\quad + \left|\Psi_{dp/r}^+\right\rangle_{x_1x_2} \otimes \left|\Psi_{dp/r}^+\right\rangle_{yy_2} + \left|\Psi_{dp/r}^-\right\rangle_{x_1x_2} \otimes \left|\Psi_{dp/r}^-\right\rangle_{y_1y_2}\Big),
\end{aligned} \tag{5.26}
$$

$$
\begin{aligned}
\left|\Phi_{dp/r}^-\right\rangle_{x_1y_1} \otimes \left|\Phi_{dp/r}^-\right\rangle_{x_2y_2} &= \frac{1}{2}\Big(\left|\Phi_{dp/r}^+\right\rangle_{x_1x_2} \otimes \left|\Phi_{dp/r}^-\right\rangle_{y_1y_2} + \left|\Phi_{dp/r}^-\right\rangle_{x_1x_2} \otimes \left|\Phi_{dp/r}^+\right\rangle_{y_1y_2} \\
&\quad + \left|\Psi_{dp/r}^+\right\rangle_{x_1x_2} \otimes \left|\Psi_{dp/r}^-\right\rangle_{y_1y_2} + \left|\Psi_{dp/r}^-\right\rangle_{x_1x_2} \otimes \left|\Psi_{dp/r}^+\right\rangle_{y_1y_2}\Big),
\end{aligned} \tag{5.27}
$$

$$\begin{aligned}\left|\Psi^{+}_{dp/r}\right\rangle_{x_1y_1} \otimes \left|\Psi^{+}_{dp/r}\right\rangle_{x_2y_2} = \frac{1}{2}\Big(&\left|\Phi^{+}_{dp/r}\right\rangle_{x_1x_2} \otimes \left|\Psi^{+}_{dp/r}\right\rangle_{y_1y_2} + \left|\Phi^{-}_{dp/r}\right\rangle_{x_1x_2} \\ &\otimes \left|\Psi^{-}_{dp/r}\right\rangle_{y_1y_2} + \left|\Psi^{+}_{dp/r}\right\rangle_{x_1x_2} \otimes \left|\Phi^{+}_{dp/r}\right\rangle_{y_1y_2} \\ &+ \left|\Psi^{-}_{dp/r}\right\rangle_{x_1x_2} \otimes \left|\Phi^{-}_{dp/r}\right\rangle_{y_1y_2}\Big),\end{aligned} \tag{5.28}$$

$$\begin{aligned}\left|\Psi^{-}_{dp/r}\right\rangle_{x_1y_1} \otimes \left|\Psi^{-}_{dp/r}\right\rangle_{x_2y_2} = \frac{1}{2}\Big(&\left|\Phi^{+}_{dp/r}\right\rangle_{x_1x_2} \otimes \left|\Psi^{-}_{dp/r}\right\rangle_{y_1y_2} + \left|\Phi^{-}_{dp/r}\right\rangle_{x_1x_2} \otimes \left|\Psi^{+}_{dp/r}\right\rangle_{y_1y_2} \\ &+ \left|\Psi^{+}_{dp/r}\right\rangle_{x_1x_2} \otimes \left|\Phi^{-}_{dp/r/r}\right\rangle_{y_1y_2} + \left|\Psi^{-}_{dp/r}\right\rangle_{x_1x_2} \otimes \left|\Phi^{+}_{dp/r}\right\rangle_{y_1y_2}\Big).\end{aligned} \tag{5.29}$$

It can be seen from entanglement swapping that there is a certain connection between the two original logical Bell states and entanglement swapping codes, and four different logical Bell states are coded, as shown below:

$$\begin{aligned}00 \rightarrow \big[&\big(|\Phi^{+}\rangle_{x_1y_1}, |\Phi^{+}\rangle_{x_2y_2}\big), \big(|\Phi^{-}\rangle_{x_1y_1}, |\Phi^{-}\rangle_{x_2y_2}\big) \\ &\big(|\Psi^{+}\rangle_{x_1y_1}, |\Psi^{+}\rangle_{x_2y_2}\big), \big(|\Psi^{-}\rangle_{x_1y_1}, |\Psi^{-}\rangle_{x_2y_2}\big)\big],\end{aligned} \tag{5.30}$$

$$\begin{aligned}01 \rightarrow \big[&\big(|\Phi^{+}\rangle_{x_1y_1}, |\Phi^{-}\rangle_{x_2y_2}\big), \big(|\Phi^{-}\rangle_{x_1y_1}, |\Phi^{+}\rangle_{x_2y_2}\big) \\ &\big(|\Psi^{+}\rangle_{x_1y_1}, |\Psi^{-}\rangle_{x_2y_2}\big), \big(|\Psi^{-}\rangle_{x_1y_1}, |\Psi^{+}\rangle_{x_2y_2}\big)\big],\end{aligned} \tag{5.31}$$

$$\begin{aligned}10 \rightarrow \big[&\big(|\Phi^{+}\rangle_{x_1y_1}, |\Psi^{+}\rangle_{x_2y_2}\big), \big(|\Phi^{-}\rangle_{x_1y_1}, |\Psi^{-}\rangle_{x_2y_2}\big) \\ &\big(|\Psi^{+}\rangle_{x_1y_1}, |\Phi^{+}\rangle_{x_2y_2}\big), \big(|\Psi^{-}\rangle_{x_1y_1}, |\Phi^{-}\rangle_{x_2y_2}\big)\big],\end{aligned} \tag{5.32}$$

$$\begin{aligned}11 \rightarrow \big[&\big(|\Phi^{+}\rangle_{x_1y_1}, |\Psi^{-}\rangle_{x_2y_2}\big), \big(|\Phi^{-}\rangle_{x_1y_1}, |\Psi^{+}\rangle_{x_2y_2}\big) \\ &\big(|\Psi^{+}\rangle_{x_1y_1}, |\Phi^{-}\rangle_{x_2y_2}\big), \big(|\Psi^{-}\rangle_{x_1y_1}, |\Phi^{+}\rangle_{x_2y_2}\big)\big].\end{aligned} \tag{5.33}$$

In a word, there is a certain connection between the two logical Bell states after entanglement swapping in the joint dephasing (rotation) noise channel. For example, two original logical Bell states $\left(|\Phi^{+}\rangle_{x_1y_1}, |\Phi^{+}\rangle_{x_2y_2}\right)$ have unique codes $'000'$, two original logical Bell states $\left(|\Phi^{+}\rangle_{x_1y_1}, |\Phi^{-}\rangle_{x_2y_2}\right)$ have unique codes $'01'$, two original logical Bell states $\left(|\Phi^{+}\rangle_{x_1y_1}, |\Psi^{+}\rangle_{x_2y_2}\right)$ have unique codes $'10'$, and two original logical Bell states $\left(|\Phi^{+}\rangle_{x_1y_1}, |\Psi^{-}\rangle_{x_2y_2}\right)$ have unique codes $'11'$.

5.3 Controllable Quantum Teleportation Authentication Model

The difference of quantum mechanics is that multiple particle systems can be entangled. For quantum system a and quantum system B, if the two subsystems are entangled, it means that some values in the properties of system a are related to property B. Even equivalent subsystems are separated in space, but some properties of the system are also related. In classical mechanics, the properties of the system are described by phase space, while in quantum mechanics, the properties of the system are described by Hilbert space. The state space of a quantum composite system is the space tensor product of N subsystems, that is, $H = \otimes_{l=l}^{n} H_l$. According to the superposition principle of states, we describe the states of the quantum composite system as follows:

$$|\Phi\rangle = \sum_{i_n} C_{i_n} |i_n\rangle, \quad (5.34)$$

where $i_n = i_1, i_2, \ldots\ldots i_n$, $|i_n\rangle = |i_1\rangle \otimes |i_2\rangle \otimes \ldots\ldots \otimes |i_n\rangle$。

Entanglement exchange is the exchange of information by measuring the basis of different entangled particles. If particles 1 and 2 are one EPR pair and particles 3 and 4 are another EPR pair, the entanglement exchange process is shown in Fig. 5.5.

Assuming that particles 1,2 and 3,4 are in the sum of quantum states $|\psi\rangle_{12}$ and $|\psi\rangle_{34}$, the state of the system is $|\Phi\rangle_{1234} = |\psi\rangle_{12} \otimes |\psi\rangle_{34}$. If the Bell state measurement is performed for particle 2 and particle 3, the quantum states of particle 1 and particle 4 will collapse after the measurement, that is, particle 1 and particle 4 are entangled and in the same state as particle 2 and particle 3.

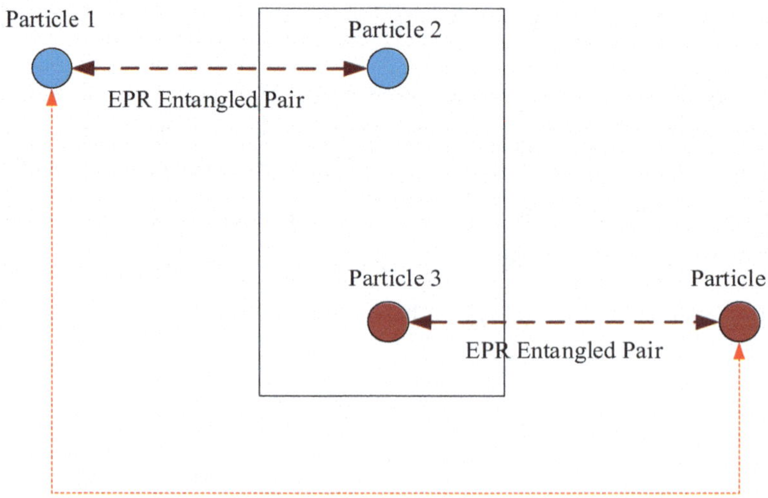

Fig. 5.5 Entanglement exchange process

5.3.1 Building a Controllable Identity Authentication Model

This section mainly verifies whether Bob's identity is legal. Therefore, we have designed a unilaterally controlled state transfer authentication model to determine Bob's identity by using quantum entanglement exchange, as shown in Fig. 5.6:

1. The three communicating parties share the entangled state $|\Phi^+\rangle$ in advance. If Alice and Charlie share the particle pair A_1 and B_1, A_2 and B_2, Bob and Charlie share the particle pair C_1 and B_3, C_2 and B_4.
2. Charlie performs Bell state measurement on his particles B_2 and B_4 and informs Alice of the measurement results. According to the entanglement exchange principle, particle A_2 owned by Alice and Bob's particle C_2 are entangled and become entangled particle pairs.
3. Bob prepares particles D and $D^{'}$ in entangled state $|\Phi^+\rangle$, sends particle D to Alice and leaves particle $D^{'}$ to himself. Bob performs Bell state measurements on particle $D^{'}$ and particle C_2 and informs Alice of the results (assuming that the measurement results are recorded as X). After Alice receives the measurement result from Bob, she performs Bell state measurement on her particles D and A_2 (the measurement result is recorded as Y).
4. Repeat the above steps until the failure rate caused by noise for t times is less than a certain critical value, and we believe that Bob's identity is legal. After Bob's identity is determined, controllable quantum information can be transmitted, thus realizing the identity authentication model of single control quantum teleportation.

In quantum networks, users A_{ij} and controllers C_i have enough EPR particle pairs $|\Phi^+\rangle$; each controller C_i and its adjacent controllers C_{i+1} also share particle pairs $|\Phi^+\rangle$, as shown in Fig. 5.7.

If we want to make the whole system communicate securely, we must first realize the identity authentication between each controller and the user, which can realize the secure communication in each controller subsystem.

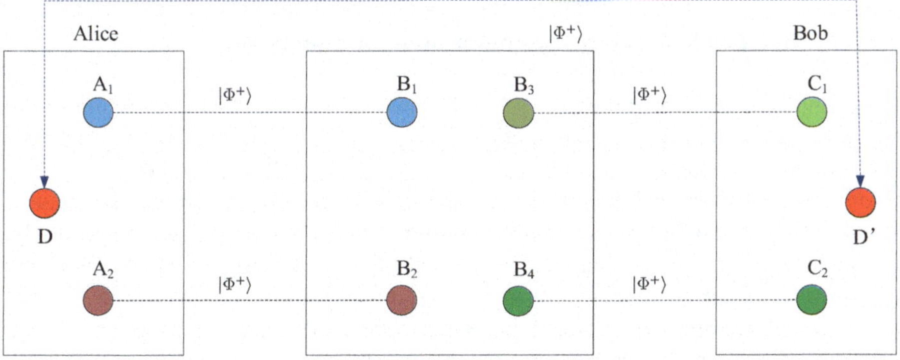

Fig. 5.6 Single control quantum teleportation identity authentication

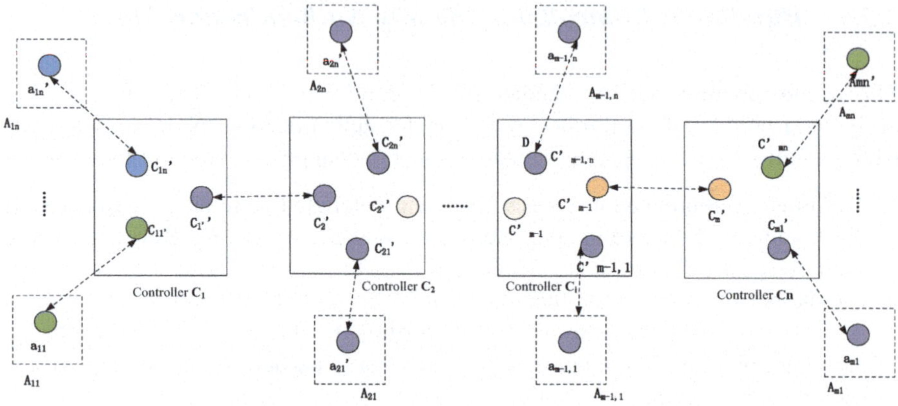

Fig. 5.7 Multicontrol quantum teleportation identity authentication

5.3.2 *Identity Authentication Process*

5.3.2.1 Judge Whether the Identity Is Legal

The identity of user A_{11} in the controller C_1 subsystem shall be judged. First of all, what A_{11} does is to send a message informing C_1 to authenticate its identity. After receiving the message from A_{11}, C_1 first prepares a pair of 1 and $1'$ EPR entangled particle pairs and sends 1 and $1'$ to A_{11} and C_1 respectively. Again, after A_{11} receives particle 1 sent by C_1, C_1 performs a series of unitary changes of Bell state measurement on particle $1'$ and C'_{11} and feeds back the changed results to A_{11}. After A_{11} receives the information from C_1, a performs Bell state measurement and a series of unitary changes on 1 and a'_{11} and records the result as X. We can know whether C_1's identity is legal through the principle of quantum entanglement exchange. Repeat the above steps n times until the failure rate is less than a certain critical value. In this way, it can be considered that the C_1 identity is legal.

5.3.2.2 Independent Safety Communication of Subsystem

If the whole system communicates securely, each subsystem must first communicate securely, that is, confirm the authentication process between the controller and the user in the subsystem. If A_{11} authenticates C_1, first A_{11} tells C_1 to authenticate him through the classical channel. When C_1 receives this message, it prepares a pair of 1 and $1'$ EPR entangled particle pairs and sends 1 and $1'$ to A_{11} and C_1, respectively. C_1 performs a series of unitary changes of Bell state measurement on particles $1'$ and C'_{11}, feeds back the changed results to A_{11}, and informs the user of the results through the classical channel. After receiving the message, the user performs Bell state measurement on his particles, records the results as y, and determines whether C_1's identity is legal according to the quantum entanglement principle. Repeat the above steps until C_1's identity is legal. After the identity of the controller is recognized, identity authentication will be conducted among users in the subsystem, that is, the controllable quantum teleportation identity authentication with C_1 as the controller is completed, which shows the security of system communication.

5.3.3 *Safety Analysis*

If there is an eavesdropper, Eve, he wants to eavesdrop the information sent by user A_{11} to user A_{m1}.

If Eve carries out entanglement measurement to attack, because entanglement measurement can only destroy the entanglement characteristics of the original particles, user A_{m1} will receive wrong information, thus increasing the bit error rate, which will eventually be found by user A_{11} in the process of eavesdropping detection.

If the eavesdropper Eve adopts the strategy of interception and retransmission, in the process of quantum teleportation, the eavesdropper intercepts particle 1 sent by user A_{m1} to user A_{11}, so that the eavesdropper has only one particle, and the user sends the entangled particle 1 to user A_{m1}, leaving particle $1'$ for himself. After receiving the particles, user A_{m1} performs Bell state measurement on the particles he owns and informs user A_{11} of the results. Then, user A_{m1} also performs Bell state measurement on the particles he owns. According to the entanglement principle, it is easy to determine whether A_{11}'s identity is legal. If the identity is legal, user A_{m1} can transmit information to user A_{11} in this quantum network. Since Eve cannot get the encoded information completely and correctly, it will only increase the bit error rate. Therefore, Eve will not get any information and will eventually be discovered by users A_{m1} and A_{11}.

For example, Eve falsifies the user A_{m1}'s identity. Since Eve cannot tell user A_{11} the measurement results, it can only guess either $|\Phi^{\pm}\rangle$ or $|\Psi^{\pm}\rangle$, and the accuracy rate is only $\frac{1}{4}$. After t times, the probability of Eve guessing correctly is $\left(\frac{1}{4}\right)^t$. If t is larger, the probability of Eve guessing correctly is smaller. Therefore, when t is large enough, the eavesdropper will be found.

5.4 Robust Quantum Teleportation Protocol Process

In this section, based on the above established secure channel, high-brightness entanglement source preparation, hyperentangled Bell state measurement and controllable identity authentication models, quantum teleportation protocols under independent different noises, and local common noises are proposed.

5.4.1 *Quantum Teleportation in Different Independent Noises*

We use a density matrix to represent quantum teleportation and qubits to represent quantum teleportation in several types of noise. Quantum teleportation is performed between Alice and Bob. If the transmitted particle is defined as the input particle, it is expressed as $|\Psi\rangle = a \mid 0\rangle + b \mid 1\rangle$, which satisfies $|a|^2 + |b|^2 = 1$. Its density matrix is expressed as follows:

$$\rho' = \mid \psi\rangle\langle\psi \mid = \begin{bmatrix} |a|^2 & ab^* \\ a*b & |b|^2 \end{bmatrix}, \tag{5.35}$$

where the subscript $*$ denotes complex conjugation. The shared quantum channel between Alice and Bob is $\left|B_1^\theta\right\rangle = \cos\theta\,|\,00\rangle + \sin\theta\,|\,11\rangle$, and its density matrix under base $\{|\,00\rangle, |\,01\rangle, |\,10\rangle, |\,11\rangle\}$ is

$$\rho'' = \left|B_1^\theta\right\rangle\left\langle B_1^\theta\right| = \begin{bmatrix} \cos^2\theta & 0 & 0 & \sin\theta\cos\theta \\ 0 & 0 & 0 & 0 \\ 0 & 0 & 0 & 0 \\ \sin\theta\cos\theta & 0 & 0 & \sin^2\theta \end{bmatrix}. \tag{5.36}$$

According to formula (5.33), the first particle is Alice's and the second particle is Bob's. When $\theta = \frac{\pi}{4}$, we get the Bell state $|\Phi^+\rangle$ and cannot set θ to a predetermined value, because it is a free parameter and can maximize the efficiency in noise-based quantum teleportation. To begin describing the protocol, we define a series of four particle orthogonal states of Alice measured by projection. We define the orthogonal state as follows:

$$\begin{aligned} \left|B_1^\phi\right\rangle &= \cos\phi\,|\,00\rangle + \sin\phi\,|\,11\rangle, \\ \left|B_2^\phi\right\rangle &= \sin\phi\,|\,00\rangle - \cos\phi\,|\,11\rangle, \\ \left|B_3^\phi\right\rangle &= \cos\phi\,|\,01\rangle + \sin\phi\,|\,10\rangle, \\ \left|B_4^\phi\right\rangle &= \sin\phi\,|\,01\rangle - \cos\phi\,|\,10\rangle. \end{aligned} \tag{5.37}$$

The four-particle Bell state can be recovered when $\phi = \frac{\pi}{4}$, and it is named $|\Phi^+\rangle, |\Phi^-\rangle, |\Psi^+\rangle$ and $|\Psi^-\rangle$. Therefore, due to a free parameter ϕ, it is selected to optimize the quantum teleportation efficiency.

Alice uses these four states to perform projection measurement, which we describe as follows:

$$P_j^\phi = \left|B_j^\phi\right\rangle\left\langle B_j^\phi\right| \quad j = 1, 2, 3, 4. \tag{5.38}$$

Before quantum operation, we first give the initial three-particle quantum state, as follows:

$$\rho = \rho' \otimes \rho''. \tag{5.39}$$

Step 1: First, Alice performs projection measurement on her two particles, that is, the input state and her shared entanglement channel. These particles will be projected onto the base $\left\{ | B_j^\phi \right\}$. After measurement, and the total quantum state will become

$$\widetilde{P}_j = \frac{P_j^\phi \rho P_j^\phi}{\mathrm{Tr}\left[P_j^\phi \rho \right]}. \tag{5.40}$$

The probability that $\widetilde{P}_j$ may occur is

$$Q_j = \mathrm{Tr}\left[P_i^\phi \rho \right]. \tag{5.41}$$

Step 2: Alice then informs Bob of the measurement result $\left| B_j^\phi \right\rangle$. After Bob knows the information, his quantum state is described as follows:

$$\widetilde{\rho}_{Bj} = \mathrm{Tr}_{12}\left[\widetilde{\rho}_j \right] = \frac{\mathrm{Tr}_{12}\left[P_j^\phi \rho P_j^\phi \right]}{Q_j}, \tag{5.42}$$

where Tr_{12} is tracking particles 1 and 2. Finally, Bob performs unitary operation U_i on his particles, so Bob's final quantum state is

$$\rho_{Bj} = U_j \widetilde{\rho}_{Bj} U_j^\dagger = U_j \frac{\mathrm{Tr}_{12}\left[P_j^\phi \rho P_j^\phi \right]}{Q_j} U_j^\dagger. \tag{5.43}$$

Bob's protocol depends not only on Alice's measurement results but also on the quantum channel used in the protocol process. For the current situation, $U_1 = I$, I is the identity matrix and $U_2 = \sigma_n$, $U_3 = \sigma_x$, $U_4 = \sigma_r \sigma_r$. At any time, the standard quantum teleportation protocol can be reconstructed when $\theta = \phi = \frac{\pi}{4}$, where, for j, $Q_j = \frac{1}{4}, \rho_{Bi} = \rho'$. For different values of θ and ϕ or when noise acts on the quantum channel, Q_i depends on the input state and average all possible independent input states, which requires estimating the efficiency of the protocol.

Previously, we have described the initial total density matrix. According to the type of each qubit in independent different noises, we have obtained the density matrix after three kinds of noises. The description is as follows:

$$p = \sum_{i=1}^{n_I} E_i(P_I)\left[\sum_{j=1}^{n_A} F_j(P_A)\left(\sum_{k=1}^{n_B} G_k(P_B)\rho G_k^\dagger\right)E_j^\dagger(P_A)\right], \tag{5.44}$$

$$E_i^\dagger(P_I) = \sum_{i=1}^{n_I}\sum_{i=1}^{n_A}\sum_{k=1}^{n_B} E_{ijk}(P_I, P_A, P_B)\rho E_{ijk}^\ddagger(P_I, P_A, P_B), \tag{5.45}$$

where $E_{ijk}(P_I, P_A, P_B) = E_i(P_I) \otimes F_j(P_A) \otimes G_k(P_B)$. Here, $E_i(P_I) = E_i(P_I) \otimes I \otimes I$, $F_j(P_A) = I \otimes F_j(P_A) \otimes I$, $G_k(P_B) = I \otimes I \otimes G_k(P_B)$ are the input qubits of Alice on the quantum channel and Bob on the quantum channel related to the noise of the Kraus operator, respectively.

In general, different noises act at different times, which is why we define the Kraus operator of P_I, P_A, P_B. By using the density matrix p, we obtain the correlation number of quantum teleportation in the noise. When $P_I = P_A = P_R = 0$ is set, a noise-free condition can be constructed.

5.4.2 *Quantum Teleportation in Local Common Mode*

5.4.2.1 Quantum Teleportation Under Local Common Pauli *x* Noise

We express the Lindblad operator in the quantum channel under $P_{nulil}\sigma_x$ noise as $L_{1,x} + L_{2,x} = \sqrt{k}\sigma_x \otimes I + I \otimes \sqrt{k}\sigma_x$. After calculation, the fidelity of quantum teleportation with the different Bell States as quantum channels is expressed as:

$$F^{co,x1} = \frac{1}{16}\left[6 - 2\cos(2\theta) + 10e^{8kt}\cos(2\theta) - (2\cos(2\theta)+2)\left(e^{8kt}-1\right)\cos(2\phi)\right], F^{co,x2} = \cos^2\phi, \tag{5.46}$$

$$F^{co,x3} = \frac{1}{16}\left[-6 + 2\cos(2\theta) + 10e^{8kt}\cos(2\theta) - (2\cos(2\theta)+2)\left(e^{8kt}+1\right)\cos(2\phi)\right],$$
$$F^{co,x4} = \frac{1}{2}[1 - \cos(2\phi)]\sin^2\phi. \tag{5.47}$$

Taking the different Bell states as quantum channels, the fidelity after experiencing $P_{nulil}\sigma_x$ noise is expressed in $F^{co,\,xy}(y = 1, 2, 3, 4)$ and substituted into the formula $F_{\text{ave}}\frac{1}{4\pi}\int_0^\pi d\phi\int_0^{2\pi} F(\phi, \theta)\sin\phi d\theta$, and we get the average fidelity under different environments as follows:

$$F_{\text{ave}}^{co,x1} = \frac{1}{3}\left(2 + e^{-8kt}\right), \tag{5.48}$$

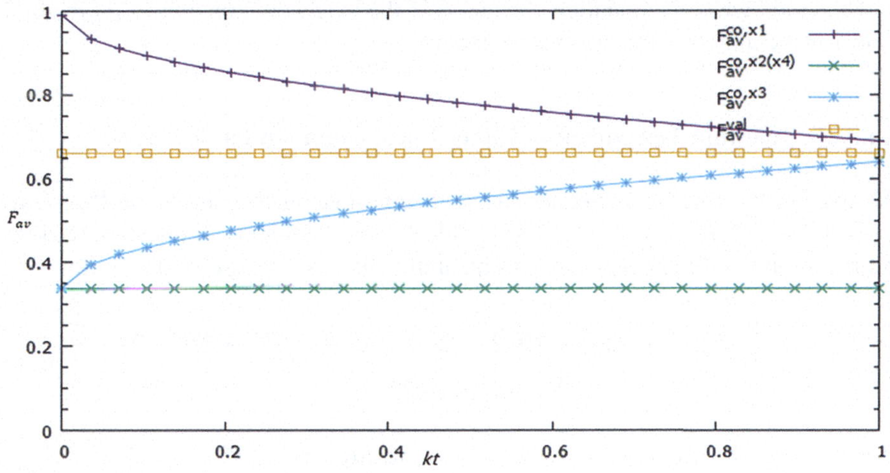

Fig. 5.8 Average fidelity under local common noise $P_{nulil}\sigma_x$

$$F_{\text{ave}}^{co,x2} = F_{\text{ave}}^{co,xt} = \frac{1}{3}, \tag{5.49}$$

$$F_{\text{ave}}^{co,x3} = \frac{1}{3}\left(2 - e^{-8kt}\right). \tag{5.50}$$

Since fidelity is the standard to measure the quality of quantum teleportation (whether it exceeds the limit value $\frac{2}{3}$), Fig. 5.8 shows the average fidelity under local common $P_{nulil}\sigma_x$ noise.

As shown in Fig. 5.8, the average fidelity of quantum teleportation under local common $P_{nulil}\sigma_x$ noise designed by us changes with the change of parameter kt and F_{anm}, and the details are as follows:

1. With Bell state 1 as the quantum channel, the fidelity changes exponentially under the local common noise $P_{nulil}\sigma_x$. When time t tends to infinity, its value gradually tends to $\frac{2}{3}$, which indicates that the Bell state 1 channel is the most suitable quantum channel at this time.
2. When Bell state 2 or Bell state 4 is used as the quantum channel, the average fidelity value under the local common noise is $\frac{1}{3}$, and because its fidelity is less than the limit value of $\frac{2}{3}$, Bell state 2 or Bell state 4 is not suitable to be used as the quantum channel at this time.
3. When Bell state 3 is used as the quantum channel, the average fidelity in the local common environment increases with the increase of parameter kt. When time t tends to infinity, its value gradually tends to $\frac{2}{3}$, but it does not exceed the limit value of $\frac{2}{3}$. Therefore, Bell state 3 is not suitable to be used as the quantum channel.

Therefore, when the quantum channel is in the local common noise $P_{nulil}\sigma_x$, Bell state 1 is the most suitable quantum channel.

5.4.2.2 Quantum Teleportation Under Local Common Pauli *y* Noise

We express the Lindblad operator in the quantum channel under noise $P_{nulil}\sigma_y$ as $L_{1,y} = \sqrt{k}\sigma_y \otimes I$, $L_{2,y} = I \otimes \sqrt{k}\sigma_y$. After calculation, the fidelity of quantum teleportation with the different Bell states as quantum channels is expressed as follows:

$$F^{in,y1} = \frac{1}{16}\left[6 + 2\cos(2\theta) + 10e^{4kt} - 2e^{4kt}\cos(2\theta) + (2\cos(2\theta) - 2)\left(e^{4kt} - 1\right)\cos(2\theta)\right], \tag{5.51}$$

$$F^{in,y2} = \frac{1}{16}\left[2 - 2\cos(2\theta) + 6e^{4kt} + 2e^{4kt}\cos(2\theta) + 6 + 2e^{4kt} - 2e^{4kt}\cos(2\theta)\cos 2\phi\right], \tag{5.52}$$

$$F^{in,y3} = \frac{1}{16}\left[-2 + 2\cos(2\theta) + 6e^{4kt} + 2e^{4kt}\cos(2\theta) - \cos(2\theta) + 6 - 2e^{4kt} + 2e^{4kt}\cos(2\theta)\cos 2\phi\right], \tag{5.53}$$

$$F^{in,y1} = \frac{1}{16}\left[-6 - 2\cos(2\theta) + 10e^{4kt} - 2e^{4kt}\cos(2\theta) + (2\cos(2\theta) - 2)\left(e^{4kt} + 1\right)\cos(2\phi)\right]. \tag{5.54}$$

Similarly, by substituting into the formula $F_{\text{ave}}\frac{1}{4\pi}\int_0^{\pi} d\phi \int_0^{2\pi} F(\phi,\theta)\sin\phi d\theta$, we can get the average fidelity under different environments as follows:

$$F_{\text{ave}}^{in,y1} = \frac{1}{3}\left(2 + e^{-4kt}\right), \tag{5.55}$$

$$F_{\text{ave}}^{in,y2} = F_{\text{ave}}^{in,y3} = \frac{1}{3}, \tag{5.56}$$

$$F_{\text{ave}}^{in,y4} = \frac{1}{3}\left(2 - e^{-4kt}\right). \tag{5.57}$$

The variation of fidelity of quantum teleportation with parameter *kt* under local common noise $P_{auli}\sigma_y$ is shown in Fig. 5.10. It can be seen from Fig. 5.9 that the average fidelity of quantum teleportation under local common noise $P_{auli}\sigma_y$ designed for us changes with the change of parameter *kt*, and $F_{an,}$ also changes. The details are as follows:

1. When Bell state 1 is used as a quantum channel for quantum teleportation, the average fidelity under local common noise $P_{auli}\sigma_y$ is better than that of the other three quantum teleportations using Bell state as the quantum channel. The average fidelity value under local common noise $P_{auli}\sigma_y$ is 1, which means that

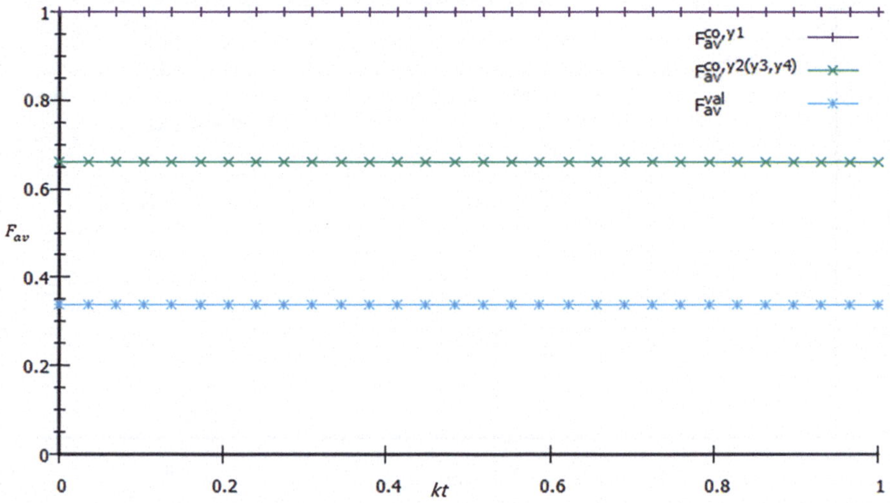

Fig. 5.9 Average fidelity under local common noise $P_{auli}\sigma_y$

the receiver can copy 100% of the quantum state to be transmitted. Bell state 1 is not affected by local common noise $P_{auli}\sigma_y$, so Bell state 1 is the most suitable quantum channel at this time.

2. When Bell state 2 or Bell state 3 is used as a quantum channel, its average fidelity value is $\frac{1}{3}$, so it is not suitable to be used as a quantum channel at this time.
3. When Bell state 4 is used as a quantum channel, its average fidelity is always $\frac{1}{3}$ under local common noise, so it is not suitable to be used as a quantum channel at this time.

Therefore, when the quantum channel is in the local common noise $P_{auli}\sigma_y$, Bell state 1 is the most suitable quantum channel.

5.4.2.3 Quantum Teleportation Under Local Common Pauli *z* Noise

We express the Lindblad operator in the quantum channel under noise $P_{auli}\sigma_z$ as $L_{1,z} = \sqrt{k}\sigma_z \otimes I$, $L_{2,z} = I \otimes \sqrt{k}\sigma_z$. After calculation, the fidelity of quantum teleportation with the different Bell states as quantum channels is expressed as follows:

$$F^{in,z1} = 1 - \frac{1}{2}\left(1 - e^{-4kt}\right)\sin^2\theta, \tag{5.58}$$

$$F^{in,z2} = 1 - \frac{1}{2}\left(1 + e^{-4kt}\right)\sin^2\theta, \tag{5.59}$$

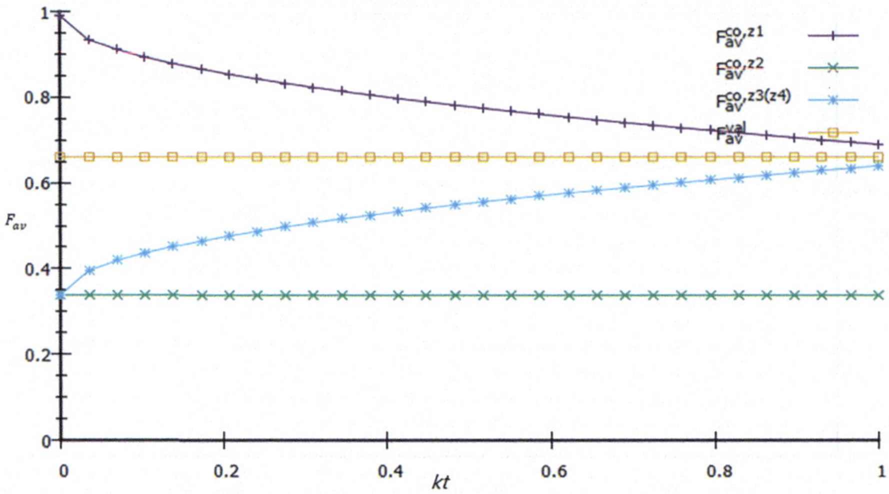

Fig. 5.10 Average fidelity under local common noise $P_{auli}\sigma_z$

$$F^{in,z3} = \frac{1}{2}\left[1 + e^{-4kt}\right) \cos(2\phi)\Big] \sin^2\theta, \tag{5.60}$$

$$F^{in,z4} = \frac{1}{2}\left[1 - e^{-4kt}\right) \cos(2\phi)\Big] \sin^2\theta. \tag{5.61}$$

Substituting into formula $F_{\text{ave}} \frac{1}{4\pi} \int_0^{\pi} d\phi \int_0^{2\pi} F(\phi, \theta) \sin\phi d\theta$, we get the average fidelity under different environments as follows:

$$F_{\text{ave}}^{in,z1} = \frac{1}{3}\left(2 + e^{-4kt}\right), \tag{5.62}$$

$$F_{\text{ave}}^{in,z2} = \frac{1}{3}\left(2 - e^{-4kt}\right), \tag{5.63}$$

$$F_{\text{ave}}^{in,z3} = F_{\text{ave}}^{in,z4} = \frac{1}{3}. \tag{5.64}$$

The variation of the average fidelity of quantum teleportation with parameter *kt* under local common noise $P_{auli}\sigma_z$ is shown in Fig. 5.10.

From the analysis of Fig. 5.10, we can see that the average fidelity of quantum teleportation under local common noise $P_{auli}\sigma_z$ designed for us changes with the change of parameter *kt*, and F_{ave} also changes. The specific situation is as follows:

1. Under local common noise $P_{auli}\sigma_z$, Bell state 1 is used as a quantum channel, and its fidelity gradually decays to $\frac{2}{3}$ in an exponential form. Therefore, Bell state 1 is the most suitable quantum channel at this time.
2. When Bell state 2 is used as a quantum channel, the average fidelity increases with the increase of parameter *kt*. When time *t* tends to infinity, the average

fidelity increases to $\frac{2}{3}$, but its limit value is less than $\frac{2}{3}$. Therefore, Bell state 2 is not suitable to be used as a quantum channel at this time.

3. When Bell state 3 or Bell state 4 is used as a quantum channel, the average fidelity value is $\frac{1}{3}$, which is not suitable for quantum channel.

In conclusion, Bell state 1 is the most suitable quantum channel when the quantum channel is in local common noise $P_{auli}\sigma_z$.

5.4.2.4 Quantum Teleportation Under Local Depolarization Noise

We express the Lindblad operator in the quantum channel under local depolarization noise as $(L_{1,\ x}, L_{2,\ x}, L_{1,\ y}, L_{2,\ y}, L_{1,\ z}, L_{2,\ z})$. After calculation, the fidelity under different noises is expressed as follows:

$$F^{in,d1} = \frac{1}{2}\left(1 + e^{-8kt}\right), \tag{5.65}$$

$$F^{in,d2} = \frac{1}{2}\left[1 + e^{-8kt}\cos(2\theta)\right], \tag{5.66}$$

$$F^{in,d3} = \frac{1}{8}e^{-8kt}\left[2\cos(2\phi) - 2 + 4e^{-8kt} - (2 + 2\cos(2\phi))\cos(2\theta)\right], \tag{5.67}$$

$$F^{in,d4} = \frac{1}{8}e^{-8kt}\left[-2\cos(2\phi) - 2 + 4e^{-8kt} + (-2 + 2\cos(2\phi))\cos(2\theta)\right]. \tag{5.68}$$

Using formula $F_{\text{ave}}\frac{1}{4\pi}\int_0^{\pi} d\phi \int_0^{2\pi} F(\phi, \theta)\sin\phi d\theta$, we get the average fidelity under different environments as follows:

$$\begin{aligned} F_{\text{ave}}^{in,d1} &= \frac{1}{2}\left(1 + e^{-8kt}\right), \\ F_{\text{ave}}^{in,d2} &= F_{\text{ave}}^{in,d3} = F_{\text{ave}}^{in,d4} = \frac{1}{2}\left(1 - \frac{1}{3}e^{-8kt}\right). \end{aligned} \tag{5.69}$$

The variation of the average fidelity of quantum teleportation with parameter *kt* under local depolarization noise is shown in Fig. 5.11.

Figure 5.11 shows the curve of the average fidelity of quantum teleportation with parameter *kt* under local depolarization noise:

1. Under local depolarization noise, the fidelity of Bell state 1 as a quantum channel decays exponentially, and finally decays to $\frac{2}{3}$, which means that it is feasible to choose this channel as a quantum channel.
2. However, for the other three Bell states, although their average fidelity decays exponentially with parameter *kt*, their limit values are less than $\frac{2}{3}$, so they are not suitable for quantum channels at this time.

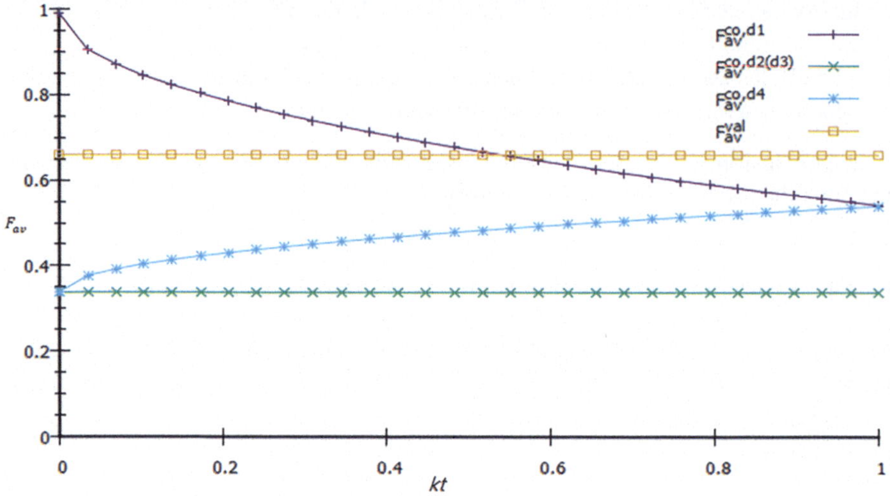

Fig. 5.11 Average fidelity under local depolarization noise

Therefore, when the quantum channel is in local depolarization noise, Bell state 1 is the most suitable quantum channel.

5.5 Efficiency Analysis of Multi-degree of Freedom Quantum Teleportation Protocol with Immune Noise

How to measure the high efficiency of the proposed quantum teleportation protocol, we use the recognized fidelity and average fidelity to evaluate the protocol.

If there is an atom A in the cavity and an atom B outside the cavity, ignoring the interaction between the two atoms, the Hamiltonian of the system is

$$H = \frac{w}{2}\sigma_z^A + \frac{w}{2}\sigma_z^B + g\left(a^\dagger \sigma_-^A + a\sigma_+^A\right) + Va^\dagger a, \tag{5.70}$$

where $a^\dagger$ is the generation algorithm of cavity photons, V is the photon frequency, σ_Z is the inversion operator of atoms, $\sigma_\pm$ is the lifting operator, g is the interaction intensity coefficient between atoms and cavities, and w is the annihilation operator of cavity photons.

If the two atoms are in an entangled state $|\psi\rangle = \cos\theta| \uparrow \downarrow\rangle + \sin\theta| \downarrow \uparrow\rangle$, where $| \uparrow\rangle$ represents the excited state and $| \downarrow\rangle$ represents the ground state, the initial state of the system is expressed by direct product as follows:

$$|\psi_0\rangle = \cos\theta|\uparrow\downarrow\rangle + \sin\theta|\downarrow\uparrow\rangle. \tag{5.71}$$

Then, the state of the system at time t is

$$|\psi'\rangle = w_1(t)|\uparrow\downarrow\rangle + w_2(t)|\downarrow\uparrow\rangle + w_3(t)|\downarrow\downarrow^{n+1}\rangle + w_4(t)|\uparrow\uparrow^{n-1}\rangle. \tag{5.72}$$

Therefore, the reduced density matrix of the two atoms at time t is

$$\rho_{AB} = \begin{bmatrix} w_4(t)w_4{*}(t) & 0 & 0 & 0 \\ 0 & w_1(t)w_1{*}(t) & w_1(t)w_2{*}(t) & 0 \\ 0 & w_2(t)w_1{*}(t) & w_2(t)w_2{*}(t) & 0 \\ 0 & 0 & 0 & w_3(t)w_3{*}(t) \end{bmatrix}. \tag{5.73}$$

5.5.1 Fidelity

Since the reference state (input state) is initially pure state, we define the fidelity as follows:

$$F_j = \mathrm{Tr}[\rho' P_{B_i}] = \langle\psi|P_{B_i}|\psi\rangle. \tag{5.74}$$

The fidelity ranges from 0 to 1, and its maximum value occurs from output state P_{B_j} to output state $|\psi\rangle$, and is zero when the two states are orthogonal. In order to consider the actual probability of occurrence of each state P_{B_j}, the average fidelity is defined as follows:

$$\overline{F} = \sum_{j=1}^{4} Q_j F_j. \tag{5.75}$$

When noise occurs or there is a nonmaximally entangled channel, $\overline{F}$ depends on the input state $\rho^{'}$. Therefore, the efficiency of the quantization protocol depends on the specific input state. The probability distribution $\rho^{'}$ of the available input state must be assumed for Alice. Here, a uniform distribution is assumed, that is, any qubit may be regarded as the input state of quantum teleportation. More specifically, without losing generality, any qubit can be written as follows:

$$|\psi\rangle = |a||0\rangle + |b|e^{ic}|1\rangle. \tag{5.76}$$

Among them, $|a|$ and $|b|$ are the absolute values of a and b, c represents the real number of the relative phase between a and b, and $|\psi\rangle$ is equivalent to $a|0\rangle + b|1\rangle$, because the difference between them is only the spherical phase without any physical significance. Since $|a|^2 + |b|^2 = 1$, $|a|^2$ and c can be selected as independent variables in order to obtain a particle c with relative phase, which is detected by state $|0\rangle$ with the possibility of $|a|^2$ and is recorded as $P'(|a|^2, c)$, where $c \subseteq [0, 2\pi]$, and $0 \le |a|^2 \le 1$.

The normalization condition $P'(|a|^2, c)$ of probability density is recorded as follows:

$$\int_0^{2\pi}\int_0^1 P'\left(|a|^2, c\right) d|a|^2 dc = 1. \tag{5.77}$$

If this certain probability distribution, $P'(|a|^2, c)$, is constant, we get

$$P'\left(|a|^2, c\right) = \frac{1}{2\pi}. \tag{5.78}$$

The average value $\langle f\rangle$ of any function $|a|^2$ and c is $\int_0^{2\pi}\int_0^1 f\left(|a|^2, c\right) P'\left(|a|^2, c\right) d|a|^2 dc$, which we define as follows:

$$\langle \overline{F}\rangle = \int_0^{2\pi}\int_0^1 \overline{F}|a|^2, c\Big) P'\left(|a|^2, c\right) d|a|^2 dc. \tag{5.79}$$

The input state mainly depends on the efficiency of quantum teleportation in noisy channels. Here, we have given both $\overline{F}\left(|a|^2, c\right)$ and $P'(|a|^2, c)$.

5.5.2 *Average Fidelity*

5.5.2.1 Average Fidelity of Single-Bit Quantum States

We express any unknown single-particle pure state as $|\Omega\rangle_n = \cos\frac{\theta}{2}|0\rangle + e^{i\phi}\sin\frac{\theta}{2}|1\rangle$ on the Bloch sphere. When azimuth angle $0 \le \phi \le 2\pi$ and polarization angle $0 \le \theta \le 2\pi$, the output state of the single-particle pure state is expressed as follows:

$$\rho' = \sum_{i=0}^{3} T_r[E_i\rho(T)]\sigma_i|\phi\rangle_n\langle\phi|\sigma_i, \tag{5.80}$$

where $|\Phi^\pm\rangle = \frac{1}{\sqrt{2}}(|00\rangle \pm |11\rangle)$, $|\Psi^\pm\rangle = \frac{1}{\sqrt{2}}(|01\rangle \pm |10\rangle)$, $E_0 = |\Psi^-\rangle\langle\Psi^-|$, and $E_1 = |\Phi^-\rangle\langle\Phi^-|$, $E_2 = |\Phi^+\rangle\langle\Phi^+|$, $E_3 = |\Psi^+\rangle\langle\Psi^+|$.

Thus, we get the output state as follows:

$$\rho = \begin{bmatrix} X & Y \\ Y & W \end{bmatrix}, \tag{5.81}$$

where

$$X = \frac{1}{2}\left(1 + \cos\left(2g\sqrt{1+nt}\right)\cos^2\phi\cos\theta + \cos\left(2g\sqrt{nt}\right)\cos\theta\sin^2\phi\right),$$
$$Y = -e^{-i\phi}\cos\left(g\sqrt{nt}\right)\cos\left(g\sqrt{1+nt}\right)\cos\phi\sin\phi\sin\theta,$$
$$W = \frac{1}{4}(2 - 2\cos\theta)\left(\cos\left(2g\sqrt{1+nt}\right)\cos^2\phi + \cos\left(2g\sqrt{nt}\right)\sin^2\phi\right).$$

The preservation or loss of quantum information in quantum teleportation is measured by fidelity, which is defined as follows:

$$F = \langle\phi|\rho|\phi\rangle. \tag{5.82}$$

When $F = 1$, it means that the quantum information is completely distorted; When $F = 0$, the input state and output state of quantum information are consistent, and the quantum information is not distorted. We know that the fidelity of single quantum state is

$$\begin{aligned} F = \frac{1}{2}&\left(1 + \cos^2\theta\left(\cos 2g\sqrt{1+nt}\right)\cos^2\phi + \cos\left(2g\sqrt{nt}\right)\sin^2\phi\right) \\ &- 2\cos\left(g\sqrt{nt}\right)\cos\left(g\sqrt{1+nt}\right)\cos\phi\sin\phi\sin^2\theta. \end{aligned} \tag{5.83}$$

For quantum teleportation, its input state is unknown and can be in any state, so the quantum transmission quality of the average fidelity of all transportable states is

$$F = \int F d\phi = \frac{1}{4\pi}\int_0^{2\pi} d\phi \int_0^{\pi} F\sin\theta d\theta. \tag{5.84}$$

From this, we get the average fidelity of the transmitted single quantum state under the teleportation:

$$\begin{aligned} F' \quad = \frac{1}{6}&\left(3 + \cos\left(2g\sqrt{1+nt}\right)\cos^2\phi - 4\cos\left(g\sqrt{nt}\right)\cos\left(2g\sqrt{1+nt}\right)\right. \\ &\left.\cos\phi\sin\phi\right) + \cos(2g\sqrt{nt})\sin^2\phi. \end{aligned} \tag{5.85}$$

Under the quantum teleportation protocol, when transferring a single qubit, the average fidelity changes with time, as shown in Fig. 5.12:

It can be seen from the analysis of Fig. 5.12 that when $\phi = \frac{\pi}{4}$, the atoms in the cavity and the atoms outside the cavity are in the maximum entangled state, and when ϕ takes different values, that is, when $\phi = \frac{\pi}{4}, \phi = \frac{\pi}{3}, \phi = \frac{2\pi}{7}$, the curve of the average fidelity with time when quantum teleportation transmits a single quantum

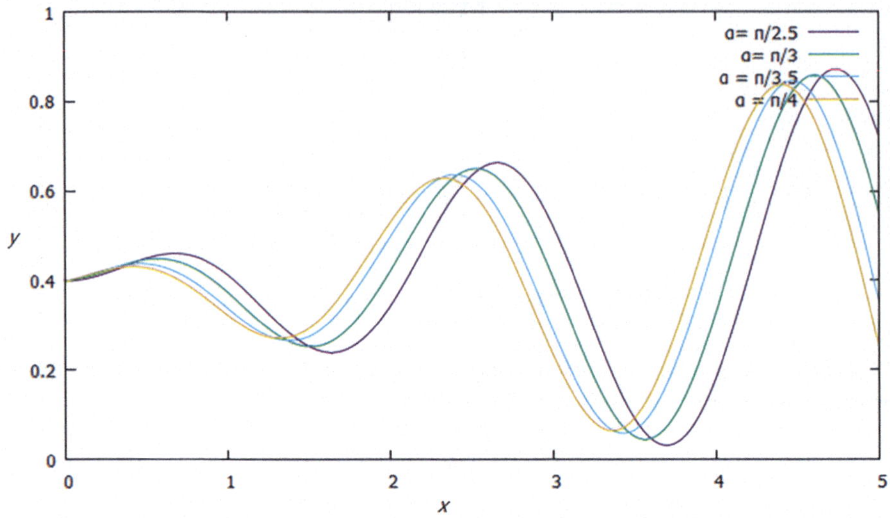

Fig. 5.12 When $g = 1$, $n = 2$, ϕ takes different values, the change of average fidelity with time when transferring a single quantum state

state. If the value of ϕ is slightly greater than $\frac{\pi}{4}$, the change of average fidelity with time is not very obvious. With the evolution of time, the channel transmission information when two atoms are in the maximum entangled state has better fidelity than that when two atoms are in the non-maximum entangled state.

When $g = 2, \theta = \frac{\pi}{4}, n$ takes different values, the change of the average fidelity of the transferred single quantum state with time under quantum teleportation is shown in Fig. 5.13.

Analysis of Fig. 5.13 shows that when ϕ takes $\frac{\pi}{12}, \frac{\pi}{8}$, and $\frac{\pi}{6}$, the average fidelity of quantum teleportation changes with time. With the gradual reduction of ϕ, the average fidelity gradually increases with time evolution. If ϕ is smaller than $\frac{\pi}{4}$, the average fidelity of channel transmission information when the two atoms are in maximum entanglement is smaller than that when the two atoms are not in maximum entanglement. Therefore, using non-entangled channel to transmit information is more advantageous.

When ϕ and coupling degree g are fixed and photon number $n = 2$, the evolution of average fidelity with time is shown in Fig. 5.14.

It can be seen from Fig. 5.14 that the average fidelity will fluctuate with time regardless of the number of photons in the cavity. The more photons, the more intense the oscillation. When there are no photons in the cavity, the average fidelity can be maximized.

At the beginning $\phi = \pi/4$, when the average photon $n = 5$, the evolution of the average fidelity with time and coupling strength is shown in Fig. 5.15.

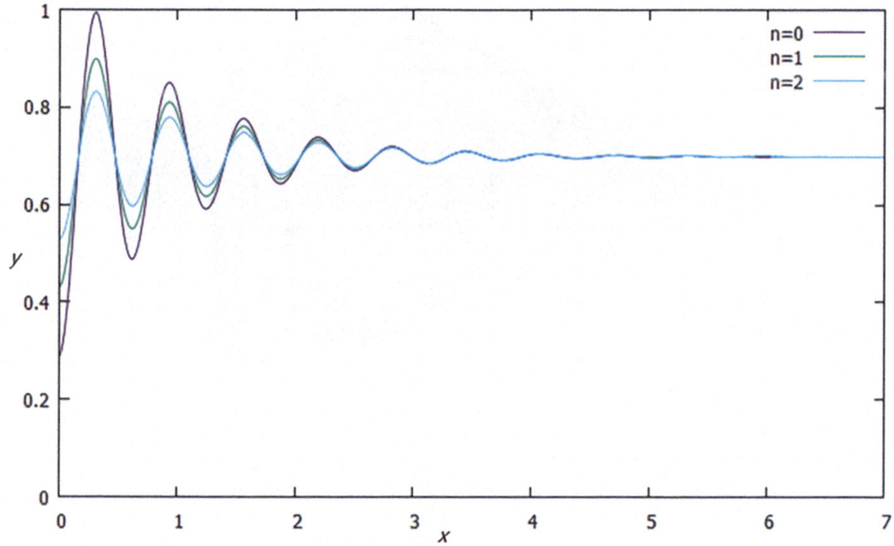

Fig. 5.13 When $g = 2, \theta = \frac{\pi}{4}$, n takes different values, the average fidelity of the transferred single quantum state changes with time

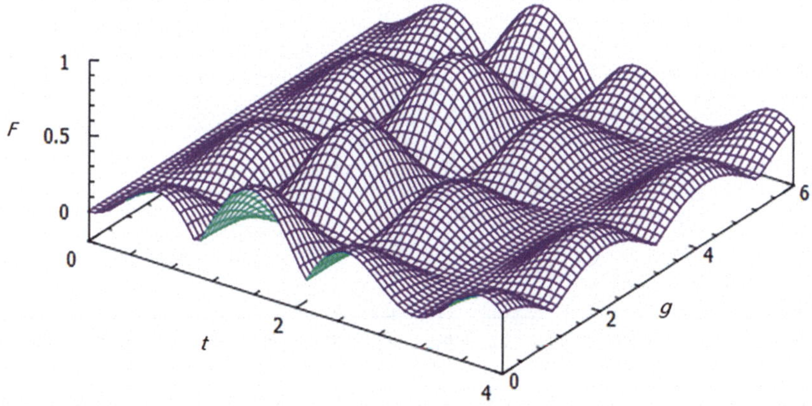

Fig. 5.14 Three-dimensional graph of average fidelity evolution with initial angle and time

5.5.2.2 Average Fidelity Under Multiple Conditions

In quantum teleportation, $E_0 = |F^-\rangle\langle F^-|, E_1 = |\beth^-\rangle\langle\beth^-|, E_2 = |\beth^+\rangle\langle\beth^+|$. Where $|F^{\pm}\rangle = \frac{1}{\sqrt{2}}(e^{i\gamma}|01\rangle \pm |10\rangle), |\beth^{\pm}\rangle = \frac{1}{\sqrt{2}}(e^{i\gamma}|00\rangle \pm |11\rangle)$, γ is the phase angle between qubits. At this time, we get the average fidelity as follows:

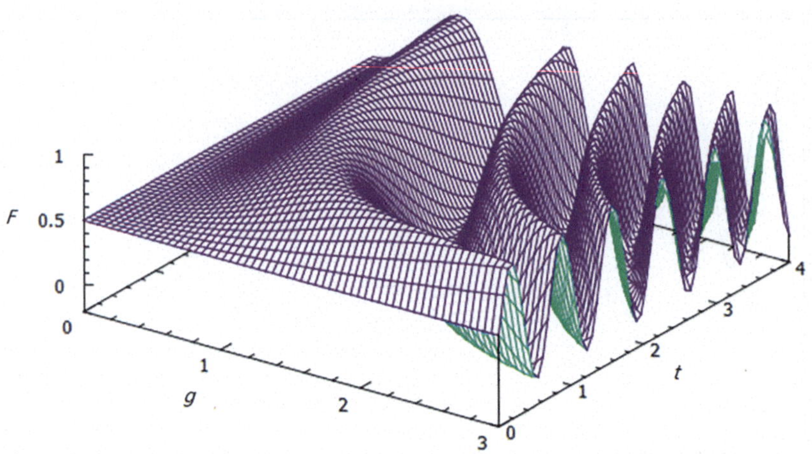

Fig. 5.15 Three-dimensional graph of average fidelity evolution with initial angle and time

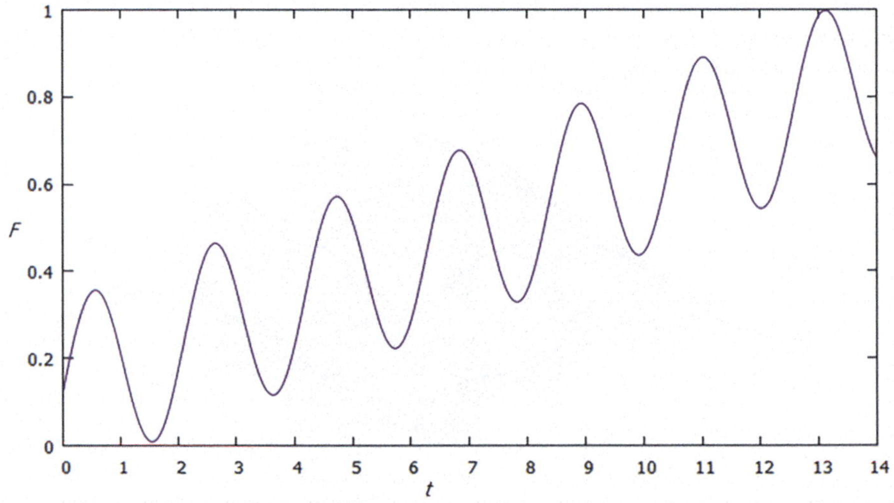

Fig. 5.16 Graph of average fidelity versus time

$$F = \frac{1}{3}\left(1 - 2\cos\left(g\sqrt{n}t\right)\cos\left(g\sqrt{1\sqrt{1+nt}}\right)\cos\phi\cos\gamma\sin\phi\right). \tag{5.86}$$

It can be seen that the average fidelity at this time is related to the coupling degree, photon number, and phase angle, as shown in Fig. 5.16.

According to the analysis of Fig. 5.16, when the coupling strength of the optical cavity is $g = 1$, the photon number is $n = 2$, the phase angle is $\gamma = 5$, and the initial angle is $\phi = \frac{\pi}{4}$, the average fidelity shows periodic oscillation with time evolution.

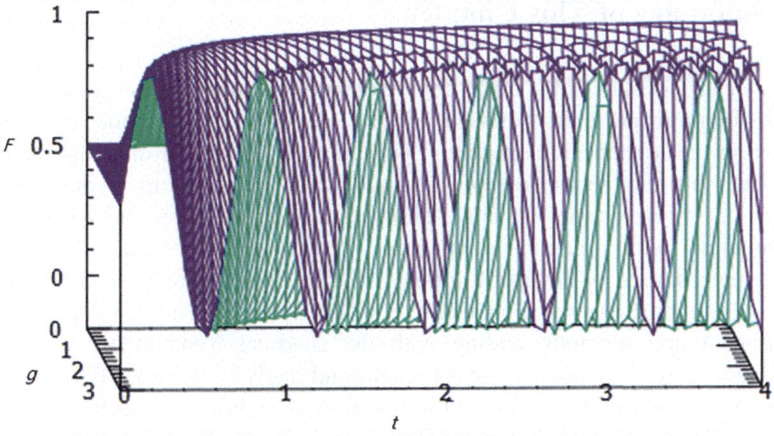

Fig. 5.17 Three-dimensional graph of the evolution of average fidelity with time

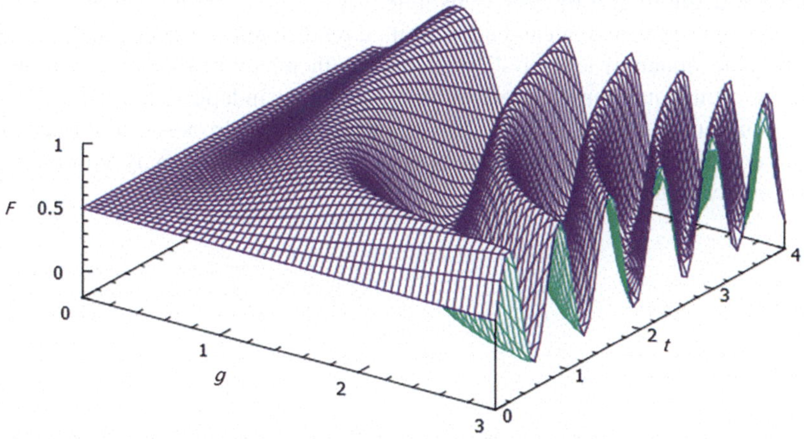

Fig. 5.18 Three-dimensional graph of the evolution of average fidelity with time

When $\phi = \frac{\pi}{4}$, $g = 5$, and $n = 1$, the average fidelity changes with γ and t when transferring a single quantum state under quantum teleportation, as shown in Fig. 5.17.

When $\phi = \frac{\pi}{4}$, $n = 2$, and $\gamma = \frac{1}{2}$quantum teleportation, the average fidelity changes with g and t when transferring a single quantum state, as shown in Fig. 5.18.

5.6 Summary of This Chapter

In this chapter, a controllable multi-degree of freedom quantum teleportation protocol with immune noise is proposed. Firstly, based on the immune noise model, considering that quantum teleportation is a linear operation of quantum states, we study the multi-degree of freedom characteristics of quantum physical system. Through the characteristics of single-photon state transmission under spin angular momentum and orbital angular momentum, we realize the preparation of quantum entanglement source with high brightness under multi-degrees of freedom. Secondly, we combine the research methods of quantum information separation and measurement and quantum coding with the classical basic theory of quantum teleportation and carry out the measurement and analysis of super-Bell states. We find the conversion relationship between spin-orbit angular momentum and super-Bell states; Then, we carry out super-Bell state super-dense coding based on graph state cascade. We quickly calculate the coherent information of various cascaded codes in the channel, the approximation value of channel capacity, and the noise tolerance of quantum information transmitted by the channel and obtain the regions where quantum information can be transmitted by different noise channels. Thirdly, a controllable quantum teleportation identity authentication model is designed to realize the robust quantum teleportation process under independent different noises and local common noises. Finally, the efficiency of the proposed multi-degree of freedom quantum teleportation protocol with immune noise is analyzed in terms of fidelity and average fidelity.

Chapter 6
Application of Fault-Tolerant Quantum Teleportation in Noisy Channel

Most of the existing quantum teleportation applications are based on the ideal quantum channel, that is, the quantum channel without any noise, and the three indicators of availability, channel utilization, and security can't be optimized at the same time under the noisy channel, which leads to the fact that these application schemes can't be applied to the field of practical quantum communication. As a necessary way to develop quantum computing and quantum network, practical quantum teleportation technology plays an important role in the fields of national defense, military affairs, finance, government affairs, quantum satellites, and long-distance communication (such as space exploration). At the same time, it will drive the rapid development of components and other upper and lower industrial chains and has become a research hotspot in the basic research field of quantum communication in recent years.

This chapter is based on the establishment of a unified framework for immune noise entangled quantum teleportation channels, the separation of quantum information between the Bell states and arbitrary states in different channels, and the construction of immune noise quantum teleportation protocols proposed in Chaps. 3–5, and we propose two specific applications to obtain real observations on quantum dialogue and quantum key distribution and expand the real applications in quantum teleportation. Designing a general fidelity quantum teleportation model and calculating an effective Bell state measurement method are the key points from theory to quantum teleportation application. As shown in Fig. 6.1, in this chapter, we designed a quantum dialogue implementation scheme for immune noise and a quantum key distribution implementation scheme based on continuous variables. The efficient, fault-tolerant, and practical channel quantum teleportation scheme is realized through two application schemes, which are of great significance to promote the application of quantum dialogue in the field of information security.

D. Li, *Quantum Teleportation Under Noise*,
https://doi.org/10.1007/978-3-031-91250-4_6

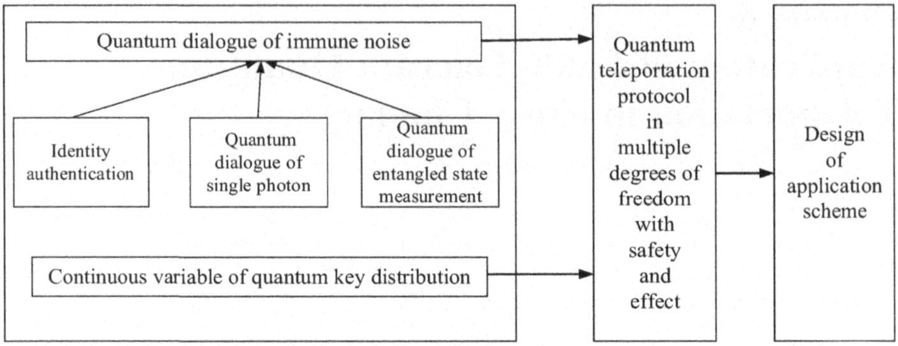

Fig. 6.1 Framework of quantum teleportation application scheme under noise

6.1 Quantum Dialogue of Immune Noise

One of the applications of quantum teleportation under collective noise is to send secret information. Quantum dialogue has become a research hotspot in the application field of quantum teleportation in recent years because of its timely and unconditional security ability of two-way secret information transmission. However, the noise problem seriously affects the security of quantum dialogue in identity authentication, information fidelity, information leakage, and so on, resulting in the inability to exchange secret information, which is a key scientific problem to be solved urgently.

To verify the identities of both sides of the dialogue and ensure that the information is not leaked under collective noise, we use decoy state and steganography technology to detect whether the identities of both sides of the dialogue are wrong and whether the dialogue channel is secure. We propose to construct a generalized entangled state by making use of the characteristics that DFS can be immune to collective noise and convert the ID codes previously shared by both sides of the dialogue into logical quantum states immune to joint dephasing noise and joint rotation noise. Random hybrid steganography is transmitted in the information sequence, which not only performs identity authentication but also eavesdropping detection, which improves the security and efficiency of the protocol.

To solve the problem that the information exchanged in the quantum dialogue protocol can't be kept true under collective noise, we propose to construct decoherent-free subspaces immune to joint dephasing noise and joint rotation noise, respectively, to ensure the accuracy of the secret information exchanged between the two sides of the dialogue.

To solve the problem of how to improve the channel capacity and long-distance control under collective noise, we use the hyperentanglement under multi-degrees of freedom as the quantum carrier and establish two-dimensional four-qubit hyperentangled states immune to noise under joint dephasing noise and joint rotation noise, respectively, to achieve dense coding. At the same time, the amount of information that can be transmitted per qubit in this model is calculated. By comparing the amount of information in other protocols, it is proved that our proposed method can achieve higher channel capacity and better remote control under collective noise.

6.1.1 Quantum Dialogue with Identity Authentication

Taking advantage of the fact that DFS is immune to collective noise, we construct a generalized entangled state, which converts the ID codes shared by both parties in advance into a logical quantum state immune to joint dephasing noise and joint rotation noise, and randomly hybridize steganography in the information sequence. Therefore, we establish secure quantum dialogue models for identity authentication and eavesdropping detection in dephasing noise and rotating noise, respectively, and establish the communication process and security analysis of the dialogue. In this way, both identity authentication and eavesdropping detection are carried out, which not only ensure that the exchange information is not leaked but also improve the security and efficiency of the protocol.

The influence of joint dephasing noise (rotation noise) on quantum states is expressed by the unitary operator:

$$U_d = \begin{bmatrix} 1 & 0 \\ 0 & e^{i\theta} \end{bmatrix}, \quad U_d = \begin{bmatrix} \cos\theta & -\sin\theta \\ \sin\theta & \cos\theta \end{bmatrix}. \tag{6.1}$$

After acting on a single particle, we can obtain $U_{dp}|0\rangle = |0\rangle$, $U_{dp}|1\rangle = e^{i\theta}|1\rangle$, $U_r|0\rangle = \cos\theta|0\rangle + \sin\theta|1\rangle$, and $U_r|1\rangle = -\sin\theta|0\rangle + \cos\theta|1\rangle$, where $|0\rangle$ is a horizontal quantum state, $|1\rangle$ is a vertical quantum state, and θ represents the noise parameters that fluctuate with time.

6.1.1.1 QD Protocol for Immunity to Noise

According to the influence of joint decoherence noise on quantum states, we construct quantum bits of immune noise as follows:

$$|L\rangle = |01\rangle_{ab}, \quad |V\rangle = |10\rangle_{ab}. \tag{6.2}$$

Among them, $|L\rangle$ and $|V\rangle$ are a set of measurement basis that represents logic bits 0 and 1, respectively, and the other measurement basis is

$$|\pm\rangle = \frac{1}{\sqrt{2}}(|L\rangle \pm |V\rangle) = \frac{1}{\sqrt{2}}(|01\rangle \pm |10\rangle) = |\Psi^{\pm}\rangle. \tag{6.3}$$

Due to $\langle +|L\rangle|^2 = |\langle + | V\rangle|^2 = |\langle - | L\rangle|^2 = |\langle - | V\rangle|^2$, $\{|L\rangle, |V\rangle\}$ and $\{|+\rangle, |-\rangle\}$ can form two unbiased bases on the joint decoherence noise. Unitary transformation can be used for conversion as follows:

$$\begin{aligned} U_0 &= I_a \otimes I_b = |L\rangle\langle L| + |V\rangle\langle V|, \\ U_1 &= (\sigma_x)_a \otimes (-i\sigma_y)_b = |V\rangle\langle L| - |L\rangle\langle V|. \end{aligned} \tag{6.4}$$

The transformation of U_1 to the measurement basis can be written as follows:

$$\begin{aligned} &U_1|L\rangle = |V\rangle, U_1|V\rangle = -|L\rangle, \\ &U_1|+\rangle = -|-\rangle, U_1|-\rangle = |+\rangle. \end{aligned} \tag{6.5}$$

Thus, a five-particle GHZ state is constructed as follows:

$$|\mathrm{GHZ}\rangle_{ab_1c_2} = \frac{1}{\sqrt{2}}(|0LL\rangle + |1VV\rangle)_{ab_1c_2}. \tag{6.6}$$

QD protocol for joint rotating noise: Because the two Bell states $|\Phi^+\rangle$ and $|\Psi^-\rangle$ are immune to noise, its logical qubits can be written as follows:

$$\begin{aligned} |L'\rangle &= |\Phi^+\rangle_{ab} = \frac{1}{\sqrt{2}}(|0\rangle|0\rangle + |1\rangle|1\rangle)_{ab}, \\ |V'\rangle &= |\Psi^-\rangle_{ab} = \frac{1}{\sqrt{2}}(|0\rangle|1\rangle - |1\rangle|0\rangle)_{ab}. \end{aligned} \tag{6.7}$$

Among them, $|L'\rangle$ and $|V'\rangle$ represent logical bits 0 and 1, $\{|L'\rangle, |V'\rangle\}$ is a group of measurement basis, and the other group of measurement basis is

$$|\pm\rangle' = \frac{1}{\sqrt{2}}(|L'\rangle \pm |V'\rangle). \tag{6.8}$$

Obviously, $|\langle +' | L'\rangle|^2 = |\langle +' | V'\rangle|^2 = |\langle -' | L'\rangle|^2 = |\langle -' | V'\rangle|^2$. Therefore, it can be known that $\{|L'\rangle, |V'\rangle\}$ and $\{|+'\rangle, |-'\rangle\}$ are two unbiased bases on the combined rotation noise, and appropriate unitary transformation is performed to convert the bases in the measurement basis to each other as follows:

$$\begin{aligned} &U_0' = I_a \otimes I_b = |L'\rangle\langle L'| + |V'\rangle\langle V'|, \\ &U_1' = I_a \otimes (-i\sigma_y)_b = |V'\rangle\langle L'| - |L'\rangle\langle V'|, \\ &I = |0\rangle\langle 0| + |1\rangle\langle 1|, \\ &-i\sigma_y = |1\rangle\langle 0| - |0\rangle\langle 1|. \end{aligned} \tag{6.9}$$

The unitary transformation of the measurement basis of U_1 is

$$U_1'|L'\rangle = |V'\rangle,\quad U_1'|V'\rangle = -|L'\rangle,\quad U_1'|+\rangle = -|-'\rangle,\quad U_1'|-\rangle = |+'\rangle. \tag{6.10}$$

Thus, a five-particle GHZ state is constructed as follows:

$$\begin{aligned} |\mathrm{GHZ}\rangle_{ab_1c_2} &= \frac{1}{\sqrt{2}}(|0L'L'\rangle + |1V'V'\rangle)_{ab_1c_2} \\ &= \frac{1}{\sqrt{2}}(|0\rangle|\Phi^+\rangle|\Phi^+\rangle + |1\rangle|\Psi^-\rangle|\Psi^-\rangle)_{ab_1c_2.} \end{aligned} \tag{6.11}$$

QD protocol that can resist two kinds of noise at the same time: if there are two kinds of noise in the quantum teleportation channel at the same time, we design a QD protocol that is immune to two kinds of collective noise at the same time according to the influence characteristics of the two kinds of collective noise on the quantum state. The logic qubits immune to both collective noises are

$$\begin{aligned} |L''\rangle &= \frac{1}{2}(|01\rangle - |10\rangle) \otimes (|01\rangle - |10\rangle) = \frac{1}{2}(|0101\rangle - |0110\rangle - |1001\rangle + |1010\rangle), \\ |V''\rangle &= \frac{1}{2\sqrt{3}}[2(|1100\rangle + |0011\rangle) - (|01\rangle + |10\rangle) \otimes (|01\rangle + |10\rangle)], \end{aligned} \tag{6.12}$$

where $\{|L''\rangle, |V''\rangle\}$ represents logical bit 0 and 1, which is a group of measurement basis, and then the other group of measurement basis is

$$|\pm''\rangle = \frac{1}{\sqrt{2}}(|L''\rangle \pm |V''\rangle). \tag{6.13}$$

And $|\langle +''|L''\rangle|^2 = |\langle +''|V''\rangle|^2 = |\langle -''|L''\rangle|^2 = |\langle -''|V''\rangle|^2$; since $\{|L''\rangle, |V''\rangle\}$ and $\{|+''\rangle, |-''\rangle\}$ are two unbiased bases on the combined rotation noise, an appropriate unitary transformation is performed to convert the bases in the measurement basis to each other as follows:

$$\begin{aligned} U_0'' &= |L''\rangle\langle L''| + |V''\rangle\langle V''|, \\ U_1'' &= |V''\rangle\langle L''| - |L''\rangle\langle V''|. \end{aligned} \tag{6.14}$$

The unitary transformation of the measurement basis of U_1'' is

$$U_1''|L''\rangle = |V''\rangle\ U_1''|V''\rangle = -|L''\rangle\ U_1''|+\rangle = -|-''\rangle\ U_1''|-\rangle = |+''\rangle. \tag{6.15}$$

Thus, a five-particle GHZ state is constructed as follows:

$$|\mathrm{GHZ}\rangle_{ab_1c_2} = \frac{1}{\sqrt{2}}(|0L''L''\rangle + |1V''V''\rangle)_{\mathrm{ab_1c_2}}. \tag{6.16}$$

6.1.1.2 Protocol Communication Process

In the joint dephasing noise channel, both sides of the dialogue should first specify that in the binary string ID of identity, if the current bit is "0," Alice needs to prepare the corresponding logical quantum state $|0\rangle_{dp} \equiv |01\rangle = |0\rangle|1\rangle, |1\rangle_{dp} \equiv |10\rangle = |1\rangle|0\rangle$. Otherwise, the logical quantum state to be prepared is

$$|+\rangle_{dp} = \frac{(|0\rangle + |1\rangle)}{\sqrt{2}}, \quad |-\rangle_{dp} = \frac{(|0\rangle - |1\rangle)}{\sqrt{2}}. \tag{6.17}$$

In the joint rotating noise channel, both sides of the dialogue should specify that in the binary string ID of identity, if the current bit is "0," Alice needs to prepare the corresponding logical quantum state $|0\rangle_r \equiv |\Phi^+\rangle, |1\rangle_r \equiv |\Psi^-\rangle$. Otherwise, the logical quantum state to be prepared is

$$|+\rangle_r = \frac{(|0\rangle_r + |1\rangle_r)}{\sqrt{2}}, \quad |-\rangle_r = \frac{(|0\rangle_r - |1\rangle_r)}{\sqrt{2}}. \tag{6.18}$$

The communication steps are as follows:

1. Alice prepared several five-particle generalized entangled state sequences according to the rules and divided them into three subsequences X, Y_1, and Y_2. Among them, sequence X is composed of x particles, sequence Y_1 is composed of y_1 particles, and sequence Y_2 is composed of y_2 particles.
2. Alice measured the particles in sequence X and recorded the measurement result as x_i and then prepared N two-particle quantum states as decoy states according to her identification code ID_X^i. If $ID_X^i = 0$, and Alice prepares the ith two-particle quantum state as $|0\rangle(|0'\rangle)$ or $|+\rangle(|++'\rangle)$; otherwise, Alice prepares the ith two-particle quantum state as $|1\rangle(|1'\rangle)$ or $|-\rangle(|-'\rangle)$. Alice inserts this decoy state into the sequence Y_1, and we get a new sequence Y_1' and send it to Bob to record the corresponding positions of all decoy states and the initial quantum states.
3. After Bob receives sequence Y_1' with the decoy particle, he informs Alice that he has received the particle sequence. Then, Alice tells Bob the position of the decoy particle in sequence Y_1', the initial quantum state, and the corresponding principle. Bob operates according to the rules and his identification code ID_Y^i, selects the corresponding measurement base pair to measure the decoy particle, and publicizes the measurement result ID_M^i.

4. Alice verifies whether $ID_Y^i = ID_X^i \bigoplus ID_M^i$ is true. If the equation is true, it proves that Bob's identity is correct and the channel is safe. Otherwise, Bob's identity is incorrect or the channel is not secure. Similarly, Bob can also verify Alice's identity and channel security. After confirming that the equation is true, communication can continue. Bob measures the remaining particles Y_1 after removing the particles in the decoy state and recording the measurement result as y_i. We know from the properties of generalized entangled states that x_i is equal to y_i.
5. Alice operates on the particles in sequence Y_2 according to the secret information X_i to get a new sequence Y_2' and disarrange the order of the particles in the new sequence Y_2'. Then she prepares N two-particle quantum states in $|0\rangle(|0'\rangle), |1\rangle(|1'\rangle), |+\rangle(|++'\rangle), |-\rangle(|--'\rangle)$ as decoy states and inserts them randomly into the new sequence Y_2' that has been disordered to get sequence Y_2'' and send it to Bob. Record the positions of all decoy states and the initial quantum states. After Bob receives sequence Y_2'', he tells Alice that he has received sequence Y_2''. Similarly, Alice announces the position and corresponding measurement basis of the decoy state, while Bob uses the measurement basis to measure the decoy state and announces the measurement results. Alice and Bob compare the measurement results with the initial decoy. If Bob's measurement results are equal to the initial state, it is proved that the channel is safe and both sides can continue to communicate. If Bob's measurement result is not equal to the initial state, the channel is proved to be unsafe, and the communication is discarded and the communication is resumed.
6. Alice informs Bob of the correct order of sequence Y_2' through the classical channel, and Bob reorders sequence Y_2', makes appropriate measurements, and announces the measurement results. Complete the quantum conversation.

In this section, in the communication between Alice and Bob (the information receiver), the quantum state of y_1 particle is taken as the signal state in the first transmission process, the decoy state is prepared according to Alice's identification code $ID_X{}^{i}$, and the decoy state is used to detect the identities of both parties and the security of the channel. In the second transmission process, the quantum state of particle y_2 was used as the signal state, and the decoy state was prepared by Alice. The quantum dialogue protocol with identity authentication proposed by us determines the decoy state before transmitting the quantum signal and adds the function of the decoy state, namely, identification. The receiver Bob needs to wait for Alice to announce the position and measurement base of the decoy state before detecting whether the communication channel is secure and the identities of the two sides of the dialogue. If the channel is secure, the communication can continue.

It can be seen from the above steps that the decoy state is mainly used to detect whether the communication channel is secure and to verify the identity of the two sides of the dialogue.

6.1.1.3 Protocol Security Analysis

In the quantum dialogue protocol with identity authentication proposed by us, Alice, the information sender, and Bob, the information receiver, have a conversation. If there is an eavesdropper, Eve, who performs unitary transformation and acts on the particles of both sides of communication, this will not only interfere with the quantum state carrying secret information but also interfere with the decoy state. As can be seen from the quantum dialogue process, half of the particles in each operation are decoy particles. If Eve performs n unitary operations and acts on n particles, it is bound to disturb the particles in the dialogue, and the probability that the interference will not be detected is

$$P = \frac{C_N^n}{C_{2N}^n} = \frac{1}{2^n}. \tag{6.19}$$

When $n \to \infty$, the probability $P \to 0$. According to the communication process of quantum communication, we can detect whether the particles of dialogue interfere.

Suppose Eve intercepts all the particles in the conversation between Alice and Bob, measures each particle, and sends a new sequence to Bob according to the measured results. The qubits in this new sequence would be randomly distributed on $\{|L\rangle, |V\rangle, |+\rangle, |-\rangle\}$(or$\{|L'\rangle, |V'\rangle, |+'\rangle, |-'\rangle\}$or $\{|L''\rangle, |V''\rangle, |+''\rangle, |-''\rangle\}$). Of course, the decoy states have N logical qubits, so half of the decoy states are distributed on $\{|L\rangle, |V\rangle\}$(or $\{|L'\rangle, |V'\rangle\}$or $\{|L''\rangle, |V''\rangle\}$).

If Eve chooses the measurement base $\{|L\rangle, |V\rangle\}(\{|L'\rangle, |V'\rangle\}, \{|L''\rangle, |V''\rangle\})$ with probability r, then the probability that this attack will be detected is $P' : P' = 1 - \left(\frac{x}{2}\right)^N$. When $N \to \infty$, the probability $P' \to 1$, so intercepted retransmission attacks will be detected.

Suppose Eve intercepts the dialogue particle between Alice and Bob and uses unitary operation U_F to make the auxiliary particle $E = \{|E_1\rangle, |E_2\rangle, \ldots\ldots, |E_n\rangle\}$ form an entangled state with the intercepted particle, thereby forming a new sequence string Y_1' and Y_2' and sending the new sequence to Bob.

For example, Eve performs a unitary transformation on the logical qubit as follows:

$$
\begin{aligned}
U_E|LE\rangle = &\quad \alpha|00\rangle|e_0e_0\rangle + \beta|01\rangle|e_0e_1\rangle + \gamma|10\rangle|e_1e_0\rangle + \delta|11\rangle|e_1e_1\rangle,\\
U_E|VE\rangle = &\quad \alpha|00\rangle|e'_n e'_n\rangle + \beta|01\rangle|e'_n e'_1\rangle + \gamma|10\rangle|e'_1 e'_n\rangle + \delta|11\rangle|e'_1 e'_1\rangle,\\
U_E|+E\rangle = &\quad \frac{1}{2}\{|\Phi^+\rangle[(\alpha|e_0e_0\rangle + \alpha|e_1e_1\rangle) + (\beta|e'_0e'_0\rangle + \beta|e'_1e'_1\rangle)]\\
&+|\Phi^-\rangle[(\alpha|e_0e_0\rangle - \alpha|e_1e_1\rangle) + (\beta|e'_0e'_n\rangle - \beta|e'_1e'_1\rangle)]\\
&+|\Psi^+\rangle[(\alpha|e_0e_1\rangle + \alpha|e_1e_0\rangle) + (\beta|e'_0e'_1\rangle + \beta|e'_1e'_0\rangle)]\\
&+|\Psi^-\rangle[(\alpha|e_0e_1\rangle - \alpha|e_1e_0\rangle) + (\beta|e'_0e'_1\rangle - \beta|e'_1e'_n\rangle)]\},\\
U_E|-E\rangle = &\quad \frac{1}{2}\{|\Phi^+\rangle[(\alpha|e_0e_0\rangle + \alpha|e_1e_1\rangle) - (\beta|e'_0e'_0\rangle + \beta|e'_1e'_1\rangle)]\\
&+|\Phi^-\rangle[(\alpha|e_0e_0\rangle - \alpha|e_1e_1\rangle) - (\beta|e'_ne'_n\rangle - \beta|e'_1e'_1\rangle)]\\
&+|\Psi^+\rangle[(\alpha|e_0e_1\rangle + \alpha|e_1e_0\rangle) - (\beta|e'_0e'_1\rangle + \beta|e'_1e'_0\rangle)]\\
&+|\Psi^-\rangle[(\alpha|e_0e_1\rangle - \alpha|e_1e_0\rangle) - (\beta|e'_ne'_1\rangle - \beta|e'_1e'_n\rangle)]\}.
\end{aligned}
\tag{6.20}
$$

If the decoy state is in $\{|L\rangle, |V\rangle\}$, Eve must make $\alpha = \beta = 0$ to prevent detection. If the decoy state is in $\{|+\rangle, |-\rangle\}$, to prevent detection, Eve must make

$$
\begin{aligned}
&\alpha|e_0e_1\rangle - \alpha|e_1e_0\rangle + \beta|e'_ne'_1\rangle - \beta|e'_1e'_n\rangle = \alpha|e_0e_1\rangle + \alpha|e_1e_0\rangle - beta|e'_0e'_1\rangle\\
&\quad - \beta|e'_1e'_0\rangle = 0.
\end{aligned}
\tag{6.21}
$$

If the above assumptions are true, Eve's entanglement measurements will not be detected.

Active attack of Eve: When Alice and Bob start quantum communication, Eve, the eavesdropper, can carry out invisible photon interception and delayed photon Trojan attack [1]. Bob can filter invisible photons by using a filter and then detect delayed photons by using the number of photons from the separator [2]. Therefore, when Bob and Alice are talking to each other, the decoy photon is randomly in the following four states $\{|0\rangle, |1\rangle, |+\rangle, |-\rangle\}$, and it is equivalent to the secure BB84 protocol [3], which can ensure the security of the conversation. The attacker Eve could be unaware of the exact location of the decoy logical qubits and the location of the measurements. If the attacker Eve executes the attack, it will be detected during the second security check. If Eve intercepts the secret order of Alice and Bob's conversation, she also has no way to know the true location and initial state of the logic qubits, so any attack will be detected.

According to the Stinespring expansion theorem [4], Eve's attack can be considered to perform a unified operation $\widehat{E}$ on a large Hilbert space $H_{AB} \otimes H_E$. Assume that Eve's auxiliary state is $|e\rangle$, so

$$
\widehat{E}|0,e\rangle = \alpha|0,e_{00}\rangle + \beta|1,e_{01}\rangle, \quad \widehat{E}|1,e\rangle = \alpha'|0,e_{10}\rangle + \beta'|1,e_{11}\rangle. \tag{6.22}
$$

There are four states of $\widehat{E}|+, e\rangle$ and $\widehat{E}|-, e\rangle$. Since $\widehat{E}$ is the overall operation, α, β, α', β' are complex and must satisfy $\widehat{E}\widehat{E}^{+} = I$; therefore, $|\alpha|^2 = |\alpha'|^2$, $|\beta|^2 = |\beta'|^2$, and then the probability of Eve attack is $e = |\beta|^2 = 1 - |\alpha|^2$.

From the point of view of information theory, the information obtained in a quantum system is not greater than Holevo's limit, that is, $\chi(\rho) = S(\rho) - \sum p_i S(\rho_i)$, where $S(\rho)$ is the von Neumann entropy, $\rho = \sum P_i \rho_i$ (the probability of preparing the quantum state of X is P through Alice). If the probability of the decoy photon passing is 1/4, then the Shannon entropy $H(\rho)$ of the decoy is $-\sum_i P_i \log_2 P_i = 2$, and then the information obtained by Eve is

$$I_E \leqslant S(\rho) - \sum_i P_i S(\rho_i) < H(P). \tag{6.23}$$

As a result, Eve cannot get a complete information about the decoy photon, and we can detect Eve's eavesdropping behavior through the interlocutor.

6.1.1.4 Protocol Comparison

The efficiency of information theory in Cabello [5] is defined as $\xi = \frac{b_s}{a_t + b_t}$, where b_s is the number of secret bits obtained by communication, q_t is the number of quantum bits, and b_t is the number of bits exchanged between Alice and Bob. The information-theoretic efficiency of the protocol is

$$\xi = \frac{b_s}{q_t + b_t} = \frac{4}{4+1} \times 100\% = 40\%. \tag{6.24}$$

In fact, the production and transmission of qubits are more complicated than classical information. Cabello's information theory efficiency formula cannot fully measure the efficiency of quantum cryptographic protocols.

In this regard, quantum bit efficiency is used to supplement effectively, which is defined as $\eta = \frac{b_n}{q_t}$, where b_n is the quantum bit used and q_t is the total quantum bit transfer. When we evaluate the efficiency of information theory, we usually combine these two parameters to evaluate it. According to the calculation, the efficiency of the proposed robust quantum dialogue protocol with immune combined rotating noise is 100%.

The proposed protocol is compared with the previous protocols 17, 18, and 19 in terms of initial quantum resources, quantum channel capacity, quantum measurement, and anti-noise. The results are shown in Table 6.1.

As can be seen from Table 6.1, in the selection of initial quantum resources, a logical Bell state was selected in [6], a logical qubit was selected in [7], and two original Bell states were selected in [8], and logical qubit and single-photon were selected in our protocol. In terms of quantum measurement, Bell measurement is adopted in literature [6, 7], and single-photon measurement is adopted in the

Table 6.1 Comparison with previous protocols

Comparative items	The protocol of [6]	The protocol of [7]	The protocol of [8]	Our protocol
Initial quantum resource	Logical Bell state	Logical qubit	Two logical Bell states	Initial Bell state and single photon
Quantum measurement	Bell state	Bell state	Single photon	Single photon
Efficiency of information theory	40%	40%	33.3%	40%
Qubit efficiency	60%	60%	66.4%	100%

protocol of literature [8]. The method of our protocol is quite simple. In terms of the efficiency of information theory, the protocol in literature [6, 7] is 40%, but it is 33.3% in literature [8]. In terms of quantum bit efficiency, this protocol is as high as 100%, but literature [6–8] only reaches 60% and 66.4%, respectively.

6.1.2 *Quantum Dialogue Based on Single Photon*

In this section, the QD protocol without information leakage is proposed by taking a single photon as the quantum resource. First, both parties in the conversation should share the rules of the conversation in advance. Single photon is used to encode and transmit secret messages. In the process of quantum teleportation, Alice (the preparator of the quantum state) needs to read (the receiver) Bob's secret message, and then Bob decodes the secret message sent by Alice according to the classical sequence of the single-photon preparation base sent by Alice. This scheme avoids information leakage and requires only a single-photon measurement.

Because $|0\rangle$ and $|1\rangle$ are the upward and downward eigenstates of $\sigma_z = |00\rangle - |11\rangle$, $|+\rangle$ and $|-\rangle$ are the upward and downward eigenstates of $\sigma_x = |01\rangle + |10\rangle$. Among them, the measurement basis of single photon is $|\pm\rangle = \frac{1}{\sqrt{2}}(|0\rangle \pm |1\rangle)$, $Z = \{|0\rangle, |1\rangle\}$, and $X = \{|+\rangle, |-\rangle\}$. There are four sets of relations as follows:

$$\begin{array}{ll} I \otimes |+\rangle = |+\rangle, & i\sigma_y \otimes |+\rangle = |-\rangle, \\ I \otimes |-\rangle = |-\rangle, & i\sigma_y \otimes |-\rangle = -|+\rangle, \\ I \otimes |1\rangle = |1\rangle, & i\sigma_y \otimes |1\rangle = |0\rangle, \\ I \otimes |0\rangle = |0\rangle, & i\sigma_y \otimes |0\rangle = -|1\rangle. \end{array} \tag{6.25}$$

From the above relationship, I and $i\sigma_y$ do not change the measurement basis, but just flip between the two states.

6.1.2.1 Communication Process

If Alice has N bits of secret message, it is denoted as $\{a_1, a_2, \ldots\ldots, a_n\}$, and if Bob has N bits of secret message, it is denoted as $\{b_1, b_2, \ldots\ldots, b_n\}$, where a_1, $b_1 \in 0$, 1, $n \in 1, 2, \ldots\ldots, N$. A single photon acts as a quantum resource in exchange for secret messages between Alice and Bob. The QD process we designed has the following steps:

1. Because Alice has N bits of secret message, she can according to her secret messages prepare a quantum state sequence Q composed of N single photon. The single photon is $\{p_1, p_2, \ldots\ldots, p_n\}$. If $a_n \neq 0$, Alice needs to prepare m photon state randomly in $\{|1\rangle, |-\rangle\}$; if $a_n = 0$, Alice needs to prepare M photon state randomly in $\{|0\rangle, |+\rangle\}$, and a classical sequence S will also be generated, marked as $\{s_1, s_2, \ldots\ldots, s_n\}$. If the preparation base of p_n is Z, then $s_n = 0$; otherwise, $s_n = 1$. To safely transmit quantum information, Alice uses the decoy-state (i.e., decoy photon) technology to prepare enough decoy photons, which are randomly in the state $\{|0\rangle, |1\rangle, |+\rangle, |-\rangle\}$ These decoy photons are randomly mixed with secret messages of N bits to form a new sequence, which is recorded as N'. Then, Alice uses block technology to transfer the new sequence N' to Bob.
2. Bob informs Alice that he has received the new sequence N', and Alice informs Bob of the location of the decoy photons in the sequence and the corresponding preparation basis. In this way, Bob measures the sequence according to Alice's preparation basis and tells Alice the measurement results. Alice judges whether the communication channel is safe according to the initial state of the decoy photon and Bob's measurement results. If the channel is secure, the communication continues, and if the channel is not secure, the communication is discarded. If the channel is secure, Bob will discard the decoy photons in the new sequence N', and the new sequence N' will become the sequence originally prepared by Alice. Bob performs a unitary operation U_{b_n} on the single-photon p_n. In this way, p_n becomes $U_{b_n}p_n$. To transmit quantum information safely, Bob also uses the decoy photon technology. Bob prepares enough decoy photons to make them randomly in any one of the states $\{|0\rangle, |1\rangle, |+\rangle, |-\rangle\}$ and mixes these decoy photons with N randomly. In this way, a new sequence will be formed, which is recorded as N''. Then Bob also transmits the new sequence to Alice through the blocking technology.
3. Perform the second security test after exchanging the roles of Alice and Bob, and the method is the same as step (2). If the channel is safe and reliable, Alice discards the decoy photons in the new sequence N'', and the new sequence N'' changes back to N. Alice measures $U_{b_n}p_n$ to decode 1 bit of Bob's b_n. Alice will prepare a new one $U_{b_n}p_n$ that has not been measured, thus forming a new N. Alice uses the decoy photon technology to prepare enough decoy photons to make them randomly in any one of the states $\{|0\rangle, |1\rangle, |+\rangle, |-\rangle\}$ and randomly mixes these decoy photons with N to form a new sequence, which is recorded as N''', and then Alice also transmits the new sequence to Bob through the blocking technology. And then perform the third safety test, using the same method as step (2).

4. Bob decodes. If the quantum channel is secure, Bob will discard the decoy photons in the new sequence N'''. Similarly, sequence N''' will become sequence N. Then, Alice publishes sequence S through the classical channel. If $s_n = 0$, Bob chooses σ_z as the measurement base to operate $U_{b_n} p_n$. If $s_n \neq 0$, Bob chooses his measurement base to operate U_{b_n}. From this, Bob can infer a_i of Alice.

6.1.2.2 Safety Analysis

If there is an eavesdropper, Eve, if he doesn't do anything but wants to get some useful information, without loss of generality, if $s_n = 0$, Alice will prepare p_n in one of the two states $\{|0\rangle, |1\rangle\}$ after receiving s_n. If p_n is in state $|0\rangle$, then $U_{b_n} p_n$ is one of the two states $\{|0\rangle, |1\rangle\}$; if p_n is in state $|1\rangle$, then $U_{b_n} p_n$ is also one of the two states $\{|0\rangle, |1\rangle\}$, so there are four possibilities for s_n. According to Shannon's information theory, it can correspond to the following:

$$-\sum_{i=1}^{4} P_i \log_2 P_i = -4 \times \frac{1}{4} \log_2 \frac{1}{4} = 2. \tag{6.26}$$

It can be seen from the formula that it can correspond to 2 bits of information, which is exactly equivalent to the number of secret messages transmitted through p_n by both sides of the communicators. From this, it can be inferred that there is no problem with information leakage in the process of quantum dialogue.

If eavesdropper Eve wants to take the initiative to attack and obtain some useful information, suppose Eve prepares a false sequence in advance and sends it to Bob after he intercepts the new sequence N'. Therefore, the result is the same as the real one when the measurement result of the false sequence is incomplete, so the probability of Eve being found is 50%. If Eve measures sequence N' after intercepting it, and then sending it back to Bob. Because Eve's measurement basis is not always the same as Alice's preparation basis for decoy photons, the probability of Eve being found is 25%. If Eve entangles its entangled photons $|e_i\rangle$ with the photons in the new sequence N' to steal some information, then

$$\widehat{E}|i\rangle \otimes |e_i\rangle = \mu|i\rangle|e_i\rangle + \nu\left|i \oplus 1\right\rangle\left|e_{i \oplus 1}\right\rangle, \tag{6.27}$$

where $|\mu|^2 + |\nu|^2 = 1$, and $\hat{\Gamma}$ is the unitary operation of eavesdropper Eve. $\langle e_i \mid e_{i \oplus 1}\rangle = 0$. Therefore, Eve will be found when both parties of the quantum dialogue perform the first security detection through base Z.

6.1.3 *Quantum Dialogue Based on Entanglement Measurement*

When photons transmit information in a time window that changes faster than the noise source, the channel noise is modeled as collective noise. This section mainly uses the entangled state of the quantum bit as the quantum initial resource to propose the QD protocol under different noise channels. Through the measurement property of the entangled state, that is, the communication parties share the initial state of the logical quantum bit in secret, in this way, the problem of information leakage is overcome, and only one photon measurement is required.

6.1.3.1 QD Protocol on Joint Decoherence Channel

If Alice owns the secret message of *N* bits recorded as $\{a_1, a_2, \ldots\ldots, a_n\}$, and Bob owns the secret message of *N* bits recorded as $\{b_1, b_2, \ldots\ldots, b_n\}$, where a_1, $b_1 \in 0, 1$, $n \in 1, 2, \ldots\ldots, N$, then the quantum dialogue on the collective decoherence noise channel is as follows:

1. Alice first prepared $M + N$ quantum bit entangled state in $\Phi\rangle_{xy}$ and then divided the quantum bit entangled state into two ordered sequences S_x and S_y, where $S_x = \{x_1, x_2, \ldots\ldots, x_n\}$ is composed of physical qubits *x* and $S_y = \{y_1, y_2, \ldots\ldots, y_n\}$ is composed of physical qubits *y*. Then, Alice sends the prepared sequence S_y to Bob in the form of block transmission, and she reserves sequence S_x for herself.
2. Bob conducts eavesdropping detection after receiving sequence S_y, choosing *M* logical qubits from the sequence S_y as sample qubits randomly, and measures each sample logical bit using *Z* base or $X \otimes X$ base:

$$X = \{|+\rangle, |-\rangle\}, |+\rangle = \frac{1}{\sqrt{2}}(|0\rangle + |1\rangle), |-\rangle = \frac{1}{\sqrt{2}}(|0\rangle - 1\rangle). \tag{6.28}$$

   Then Bob informs Alice of the measurement basis and location of these sample logical qubits, and Alice chooses the appropriate measurement basis to measure the sample logical qubits in S_x. If there is an eavesdropper, the communication will be discarded. If there is no eavesdropper, the communication will continue.
3. Alice and Bob discard the sample logical qubits in S_r and S_z, so that the original sequence of S_x and S_y will become S'_x and S'_y. Record as $S'_x = \{x_1, x_2, \ldots\ldots, x_n\}, S'_y = \{y_1, y_2, \ldots\ldots, y_n\}$. Alice and Bob measure with bases *Z*, respectively. According to the correlation principle of entanglement measurement, Alice and Bob can infer each other's test results. Bob prepares a new sequence y_n according to his test results, and then Bob performs the unitary operation U^{i_n} on y_n. In this way, the new sequence φ_n will become $U^{i_n} y_n$. Bob uses the decoy photon technology to prepare enough decoy logic qubits, which

are in any state of $\{|0\rangle, |1\rangle, |+\rangle, |-\rangle\}$, and randomly inserts the prepared decoy logic qubits into sequence S'_y, so that sequence S'_y becomes sequence S''_y. Bob sends the new sequence S''_y to Alice in the form of block technology. Alice informs Bob that the sequence has been received, and Bob informs Alice of the preparation basis and location of the decoy logic qubits. Alice measures the decoy logic qubits according to the preparation basis told by Bob and tells Bob the measurement results. Bob determines whether the channel is safe according to Alice's measurement results and the initial state of the decoy logic qubits. If the channel is not safe, the communication will be terminated. If the channel is safe, the communication will continue.

4. Alice discards the decoy logic qubits in sequence S''_y, so sequence S''_y becomes S'_y. Alice performs a logical unitary operation on sequence $U^{i_n}y_n$, so $U^{i_n}y_n$ becomes $U^{a_n}U^{i_n}y_n$. Then, Alice uses base Z to measure $U^{a_n}U^{i_n}y_n$ and publishes the measurement results. According to the unitary operation, the measurement result base Z, and the results published by Alice, Bob can know Alice's secret message a_n. Accordingly, Alice can also infer Bob's secret bit message b_n.

6.1.3.2 QD Protocol on Joint Rotating Channel

1. Alice prepares $M + N$ quantum bits entangled state in $|\Phi\rangle_{x\eta}$ and then divides the quantum bit entangled state into two ordered sequences S_x and S_y, where $S_x = \{x_1, x_2, \ldots\ldots, x_n\}$ is composed of physical qubits x and $S_y = \{y_1, y_2, \ldots\ldots, y_n\}$ is composed of physical qubits y. Then, Alice sends the prepared sequence S_y to Bob in the form of block transmission, and she reserves sequence S_x for herself.
2. Bob conducts eavesdropping detection after receiving sequence S_y, randomly chooses M logical qubits from sequence S_y as sample qubits, and measures each sample logical bit using Z base or $Y \otimes Y$ base:

$$Y = \{|+y\rangle, |-y\rangle\}, |+y\rangle = \frac{1}{\sqrt{2}}(|0\rangle + i|1\rangle), |-y\rangle = \frac{1}{\sqrt{2}}(|0\rangle - i|1\rangle). \tag{6.29}$$

Alice uses Z bases or Y bases to measure sample qubits in S_x. Then, Alice and Bob use Z to measure sequences S_x and S_y, respectively. According to the measurement results, Bob prepares a new sequence S_y and then performs a logical unitary operation on the new sequence S_y, so we get $U^{i_n}y_n$; the logical qubits prepared by Bob are randomly in any of the states $\{|0\rangle, |1\rangle, |+\rangle, |-\rangle\}$.

3. Alice informs Bob that he has received S''_y, and Bob informs Alice of the preparation basis and location of the decoy logic qubits. Alice uses Bob's preparation basis to measure the decoy logic qubits and inform Bob of the measurement results. Bob determines whether the channel is safe by comparing Alice's measurement results with the initial state of the decoy logic qubits. If the channel is not safe, the communication will be terminated, and if the channel is safe, the communication continues. Alice performs a unitary operation on logical

qubits $U^{i_n}y_n$ to obtain $U^{a_n}U^{i_n}y_n$. Then, Alice uses the Z base to measure $U^{a_n}U^{i_n}y_n$, and publishes the measurement results to Bob, and then uses U^{i_n} and the measurement results of S_y to decode b_n.

6.2 Quantum Key Distribution with Continuous Variables

Another application of quantum teleportation is quantum key distribution. Arbitrary single-mode quantum states and dual-mode quantum states can carry out quantum long-distance transmission and can also realize long-distance transmission of coherent states as long as the degree of entanglement is high, but they have shortcomings. Because of the shortcomings of the existing key distribution protocols, we propose continuous-variable quantum key distribution, which mainly means that the sender transmits the predetermined key to the information receiver through the public channel and with the help of the entanglement exchange principle of entangled coherent states. Using the characteristics of three quantum GHZ states and W states, there are four GHZ states in Hilbert space. If Alice prepares a large number of three quantum polarized GHZ states $|\Phi_A\rangle$ whose number must be greater than the length of the key, it is divided into two sequences A and B. A sequence is composed of $\{\mathrm{a}_1, a_2, \ldots\ldots, a_n\}$ and B sequence is composed of $\{\mathrm{b}_1, b_2, \ldots\ldots, b_n\}$.

The beam can produce two-component amplitudes X and phases P, which can be expressed by operators as follows:

$$X = a^\dagger + a, \quad P = i\left(a^\dagger - a\right), \tag{6.30}$$

where $a^\dagger$ is the production operator, a is the annihilation operator, $[a^\dagger, a] = 1$, and $| X, P] = 2i$.

Two-mode squeezed vacuum states are mainly prepared by optical parametric amplification of two vacuum states. Assuming that the two vacuum states are a_1 and a_2, the output state is

$$\begin{aligned} a_1' &= S^\dagger(r) a_1 S(r) = a_1 \cos h(r) + a_2^\dagger \sin h(r), \\ a_2' &= S^\dagger(r) a_2 S(r) = a_2 \cos h(r) + a_1^\dagger \sin h(r). \end{aligned} \tag{6.31}$$

Therefore, the amplitude and phase of a_1 and a_2 are

$$\begin{aligned} X_1' &= X_1 \cos h(r) + X_2 \sin h(r), \\ P_1' &= P_1 \cos h(r) - P_2 \sin h(r), \\ X_2' &= X_2 \cos h(r) + X_1 \sin h(r), \\ P_2' &= P_2 \cos h(r) - P_1 \sin h(r). \end{aligned} \tag{6.32}$$

Thus, the phase and quadrature amplitude correlation variances of the two outputs can be calculated as follows:

$$\begin{aligned}\left[\triangle\left(X_1' - X_2'\right)\right]^2 &= \left[\triangle\left(P^{1\prime} + P^{2\prime}\right)\right]^2 = 2e^{-2r},\\ \left[\triangle\left(X_1' + X_0'\right)\right]^2 &= \left[\triangle\left(P^{1\prime} - P^{2\prime}\right)\right]^2 = 2e^{2r}.\end{aligned} \tag{6.33}$$

When $r \to +\infty$, the relevance of a_1 and a_2 is

$$\lim_{r\to+\infty} X_1' = X_2', \lim_{r\to+\infty} P_1' = -P_2'. \tag{6.34}$$

Knowning that the entanglement degree of the two-mode squeezed state is K, we describe it as follows:

$$F = \left[\triangle\left(X_1' - K_1X_2'\right)\right]^2 \times \left[\triangle\left(P_1' + K_2P_2'\right)\right]^2. \tag{6.35}$$

When $k_1 = \frac{x_1'x_2'}{x_2'^2}$, $k_2 = -\frac{p_1'p_2'}{p_2'^2}$, we can get

$$F = \frac{4\sigma^4}{e^{2r} + e^{-2r}}. \tag{6.36}$$

When $r \to +\infty$, $F \to 0$, if the entanglement is destroyed, r will become smaller, which will cause F (the entanglement between two-mode squeezed states) to increase.

When Alice and Bob share entangled states a_1 and a_2, Alice prepares coherent states $|x + ip\rangle_1$ and performs joint Bell basis measurements with a_1', so we get

$$\begin{aligned}x_u &= \frac{1}{\sqrt{2}}\left(x_1 - x_1'\right),\\ p_u &= \frac{1}{\sqrt{2}}\left(p_1 + p_1'\right).\end{aligned} \tag{6.37}$$

If Bob receives the signal,

$$\begin{aligned}x_B &= x' + \sqrt{2}x_u - \sqrt{2}x_u = x_1 - \left(x_1' - x_2'\right) - \sqrt{2}x_u,\\ p_B &= p' + \sqrt{2}p_u - \sqrt{2}p_u = p_1 + \left(p_1' + p_2'\right) - \sqrt{2}p_u.\end{aligned} \tag{6.38}$$

Bob performs proper unitary transformation after receiving Alice's measurement results to obtain

$$x'_B = x_B + \sqrt{2}x_u = x_1 - (x'_1 - x'_2),$$
$$p'_B = p_B + \sqrt{2}p_u = p_1 + (p'_1 + p'_2). \tag{6.39}$$

So Alice and Bob get two sequences to pass the key. If the entanglement degree of a_1 and a_2 is low, we can't get accurate results. When Alice passes the key, if $x_1 = p_1$, sequence x_1 can be obtained regardless of Bob's choice of phase or amplitude measurement, so the key distribution can also be realized.

6.2.1 Key Distribution Process

If Alice wants to pass the quantum key to Bob, the steps are as follows:

1. The two-mode squeezing operator is expressed as $S(r)$, and the vacuum state is expressed as $|00\rangle_{xy}$, acting $S(r)$ on $|00\rangle_{xy}$, and then produce entangled optical modes, which are expressed as a_x and a_y. Alice calculates the entanglement degree F between a_x and a_y through the random time gap and then sends the optical mode to Bob.
2. If there is no eavesdropper in the communication channel after Bob receives the optical mode $a_z(a_z = a_y)$, Alice returns the time threshold of the transmission, and Bob measures the phase or amplitude according to the time threshold to calculate the degree of entanglement F'. If $F' > F$, it proves the existence of the eavesdropper, thus giving up the communication. If $F' = F$, there is no eavesdropper, and the communication continues.
3. Alice first encodes the discrete information according to the graph state basis, divides the interval, and then sends the information. Suppose Alice divides the coordinate interval into $(-\infty, a_1)[a_1, a_2)$, …, $[a_{n-1}, a_n)[a_n, +\infty)$; if the sequence 01001 is to be sent, the corresponding interval is $[a_{k-1}, a_k)$, so the random variables generated are a, b, and c. If the bit shift operator is expressed by W, that is, $\alpha_1 = (a + b) + i(a + b)$, applying W to the vacuum state $|0\rangle_1$, it can produce coherent states $|x\rangle_1$. Applying W to vacuum state $|0\rangle'_1$, it can produce coherent states $|x\rangle'_1$, that is, decoy states. In a random time gap, a quantum mixing state r is generated by inserting $|x\rangle'_1$ into $|x\rangle_1$. Then, Alice will perform joint Bell state measurement in r and a_r to obtain the corresponding phase difference P_u and amplitude X_u. After measurement, Alice will inform Bob of the measurement results through the classical channel.
4. Bob performs a unitary transformation on a_z after receiving the measurement results, recording as D, and then measures the amplitude or phase according to the corresponding measurement basis, to obtain sequence ω. Then, Alice informs Bob of the time gap of decoy state $|x\rangle'_1$, and Bob publishes the measurement results of the time gap. If the results are inconsistent, the two sides will give up this communication. If the comparison results are consistent, the communication will continue. Bob removes the remaining measurement results from sequence ω,

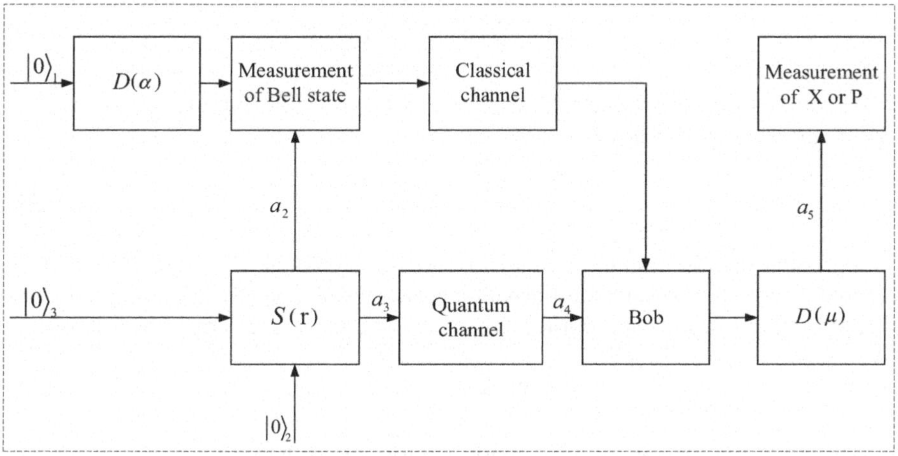

Fig. 6.2 Quantum key distribution protocol model of continuous variables

and Alice decodes them in the way of graph-based coding, so we get $a(a \subseteq [a_{k-1}, a_k)$. Finally, Bob gets the bit sequence 01001.

In short, the optical film generates the entanglement, the degree of entanglement calculated by the time gap, and another degree of entanglement obtained by the amplitude or phase measurement. The purpose of these steps is to detect the presence of eavesdroppers and share the quantum entangled states of continuous variables safely between Alice and Bob. The sequence is sent to generate a coherent state and decoy state. After mixing them into a mixed state, the Bell state measurement is performed to obtain the phase difference and amplitude. The receiver performs a unitary transformation to obtain the sequence. The results are compared according to the measurement results of the time gap, and the bit sequence is obtained after decoding. These steps are the key sending, key receiving, and generation stages. The purpose is that the key is random except for the sender. The purpose of random translation is that the key is not equivalent to the result. Therefore, the key can be obtained only after the channel is secure and authenticated.

Therefore, the quantum key distribution protocol model of continuous variables is shown in Fig. 6.2.

6.2.2 Safety Analysis

In the continuous-variable quantum key distribution protocol designed by us, Bob selects n or p for random measurement, and the probability is 1/2, and then

$$\Delta\Upsilon = \frac{1}{2}\left(\Delta\Upsilon_x + \Delta\Upsilon_p\right). \tag{6.40}$$

Because α or p has symmetry, then $\Delta\Upsilon_x = \Delta\Upsilon_p$, so the rate of information transfer can be abbreviated as follows:

$$\Delta\Upsilon = \Delta\Upsilon_x. \tag{6.41}$$

Assuming that the amount of mutual information between Alice and Bob is $\Upsilon(x_B, x_A)$, and the amount of mutual information between Bob and Eve is $\Upsilon(x_B, x_E)$, the rate of information transmission during reverse coordination is described as follows:

$$\Delta\Upsilon_x = \Upsilon(x_B, x_A) - \Upsilon(x_B, x_E). \tag{6.42}$$

According to the analysis of Shannon's information theory,

$$\begin{aligned} I(x_B, x_A) &= \frac{1}{2}\log_2\left(\frac{V_B}{V_{A/B}}\right), \\ I(x_B, x_E) &= \frac{1}{2}\log_2\left(\frac{V_E}{V_{E/B}}\right), \end{aligned} \tag{6.43}$$

where Alice's variance about Bob is expressed in $V_{A/B}$ and Eve's variance about Bob is expressed in $V_{E/B}$. Therefore, it is limited for Alice and Eve to obtain Bob's information at the same time, and it satisfies the Heisenberg uncertainty relationship. When $V_{A/B}$ gets the minimum value, the lower limit of the maximum information that Eve can obtain is V_E^x/B°.

In this protocol, both parties have conducted a quantum signal transmission in the quantum channel. If Eve adopts the interception and retransmission strategy to obtain the signal sent by Alice, then randomly select P or X to measure and then prepare the corresponding quantum state according to the measurement results and send it to Bob. According to the quantum uncertainty principle, Eve will introduce additional noise, which will cause the entanglement degree to change and eventually be discovered by both sides of the communication. Eve may also intercept some of the information sent by Alice. According to the quantum non-cloning principle, Eve intercepts the beam sent by Alice and lets it pass through the spectroscope. The projection coefficient η is equal to the transmission efficiency of the channel. Except for the previously intercepted part, the rest will be transmitted to Bob. We record the coherent state prepared by Alice as $a_1 = \varpi_1 + a$, which is expressed in terms of x components as $x_1 = x_1 + x$; therefore, the variance is $V_1^x = \sigma^2 + V$. If eavesdropper Eve intercepts the beam emitted by Alice and passes through the spectroscope, we get

$$\begin{aligned} a_z &= \sqrt{x}a_z + \sqrt{1-x}a_n, \\ a'_F &= \sqrt{x}a_n - \sqrt{1-x}a_z, \end{aligned} \tag{6.44}$$

where a_n is the noise of the quantum channel. After Alice publishes the information, Bob and Eve perform unitary operations, respectively, as shown below:

$$\begin{aligned} a_m &= a_z + \sqrt{x}\mu, \\ a_E &= a'_{F'} - \sqrt{(1-x)\mu}. \end{aligned} \tag{6.45}$$

The definition of variance $V_{A/B}$ is described as follows:

$$V_{A/B} = a_w^2 - \frac{(a \cdot a_w)^2}{a^2}. \tag{6.46}$$

Thus, the r component is

$$\begin{aligned} x_5 &= x_4 + \sqrt{\eta}(x_1 - x_2) \\ &= \sqrt{\eta}x_3 + \sqrt{(1-\eta)}x_N + \sqrt{\eta}(x_1 - x_2), \end{aligned} \tag{6.47}$$

$$\begin{aligned} V_{A/B} &= 1 - \eta + 2\eta(\cos h^2(r) + \sin h^2(r)) - 4\eta\cos \cosh(r) \sin h(r) + \eta V_1 - \eta V \\ &= 1 - \eta + \eta V + 2\eta e^{-2r}. \end{aligned} \tag{6.48}$$

The variance of the correlation state on the two components is consistent. The squeezed state can compress the components to infinity. When x_1 prepared by Alice is a squeezed state, then $V_1 = \sigma^2 + V$. According to the properties of the squeezed state, we get

$$V^p_{A/B} = 1 - \eta + \frac{\eta}{V^x_1} + 2\eta e^{-2r}, \tag{6.49}$$

$$V^x_{A/B} = \left(1 - \eta + \frac{\eta}{V^x_1} + 2\eta e^{-2r}\right)^{-1}, \tag{6.50}$$

$$\Delta I \geq -\frac{1}{2}\log_2\left(1 - \eta + \frac{\eta}{V_1} + 2\eta e^{-2r}\right)(1 - \eta + \eta V_N + 2\eta e^{-2r}). \tag{6.51}$$

The change curve of the information transmission rate is shown in Fig. 6.3.

According to Fig. 6.3, when $r = 0$, there is no entanglement, and when $\Delta I < 0$, the protocol we designed cannot transfer the key. Therefore, it can be seen that the entanglement property of quantum is the basis of quantum remote communication. If the entangled pair between Alice and Bob is destroyed, Bob can't receive the key information. When $r > 0.58$, the amount of information is about zero; when $r = 1$, the amount of information is greater than zero. The key can be transmitted safely

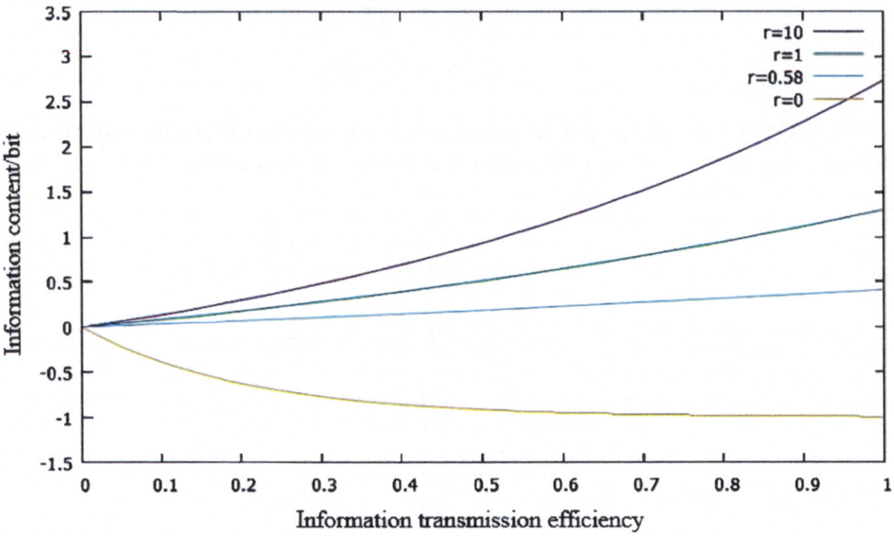

Fig. 6.3 Change curve of information transmission rate

when the channel transmission rate is greater than zero. The amount of mutual information between Alice and Bob increases with the increase of entanglement. When $r = 10$, the channel transmission efficiency is $\eta = 0.75$, and the key transmission rate is 1 bit/s, and when the channel transmission efficiency is $\eta = 0.95$, the key transmission rate reaches 2 bit/s. With the increase of channel transmission rate, the speed of continuous-variable quantum key distribution also increases, and its effect is better than discrete quantum key distribution.

6.3 Summary of This Chapter

Based on the previous chapters, aiming at the practical bottleneck of quantum teleportation, this chapter designs two typical applications: quantum dialogue with immune noise and quantum key distribution with continuous variables. Firstly, a dialogue mechanism using decoy state and steganography is proposed to detect whether the identity of both sides of the dialogue is wrong and whether the dialogue channel is secure. The single-photon and generalized entangled states are constructed to convert the ID code shared by both sides into a logical quantum state immune to joint dephasing noise and joint rotation noise. The random hybrid steganography is transmitted in the information sequence, which performs not only identity authentication but also eavesdropping detection. The communication parties only need to carry out the single-photon measurement to avoid information leakage and improve the robustness of the quantum dialogue process. Secondly, a continuous-variable quantum key distribution protocol is proposed. The sender transmits the predetermined key to the information receiver through the public channel. In the key generation phase, the decoy state and translation are randomly inserted into the sequence to ensure the security of quantum key distribution. In a word, the dialogue protocol immune to noise and quantum key distribution proposed by us has high robustness, more potential, and fault tolerance under noise and opens up a new application field for further promoting the practical quantum teleportation technology.

References

1. H. Ryszard, H. Paweł, H. Michał, H. Karol, Quantum entanglement. Rev. Mod. Phys. **81**(1), 865–942 (2009)
2. Z. Margherita, K. Tanjung, P. Tomasz, B. Somshubhro, B. Anindita, D. Prasenjit, H. Saronath, M. Kavan, P. Mauro, Excessive distribution of quantum entanglement. Phys. Rev. **93**(1), 012305 (2016)
3. P.C. Obando, F.M. Paula, M.S. Sarandy, Trace-distance correlations for X, states and the emergence of the pointer basis in Markovian and non-Markovian regimes. Phys. Rev. A **92**(3), 676–677 (2015)
4. V. Špička, T.M. Nieuwenhuizen, P.D. Keefe, Physics at the FQMT'11 conference. Phys. Scr. **151**(1), 014001 (2012)
5. 喻思娈 (S. Yu), 首颗量子卫星"墨子号"在轨交付 (First Quantum Satellite 'Micius' Delivered for Orbital Operation). 人民日报, 2017年1月19日 (People's Daily, January 19, 2017)
6. 王蔚蔚 (W. Wang), 金融信息量子通信验证网开通 (Financial Information Quantum Communication Verification Network Launched). 合肥日报, 2012年2月23日 (Hefei Daily, February 23, 2012)
7. 宗合,宗媛 (H. Zong, Y. Zong), 合肥城域量子通信试验示范网全网开通仪式举行 (Full Network Launch Ceremony Held for Hefei Metropolitan Quantum Communication Trial Demonstration Network). 合肥日报, 2012年3月31日 (Hefei Daily, March 31, 2012)
8. 刘诗瑶 (S. Liu), 我国成功发射世界首颗量子卫星 (China Successfully Launches World's First Quantum Satellite). 人民日报, 2016年8月16日 (People's Daily, August 16, 2016)

Chapter 7
Summary and Outlook

The key scientific problem of this book is the challenge to traditional cryptographic algorithms brought by privacy information leakage and eavesdropping network information security. Starting from the new perspective of entanglement death and multi-degree of freedom characteristics of quantum systems, the unified entanglement quantum teleportation channel framework theory and quantum teleportation protocol are proposed, which have high academic research value and great breakthrough space. This book focuses on the key technologies and applications of quantum teleportation under noise. Guided by the security-oriented information security communication, then the characteristics analysis and theoretical modeling of quantum teleportation are carried out. Based on the application of security in noisy channels, the theoretical model and key methods are evaluated and verified. The concept of interdisciplinary is introduced to study the entanglement evolution and channel capacity in local independent and local common modes. It effectively solves the problems of achieving high-capacity, low-cost, easily prepared, and highly safe quantum teleportation under noise.

7.1 Summary of Research Work in This Book

Quantum teleportation is the key technology of quantum communication. It mainly focuses on designing and proposing a scheme of quantum teleportation that meets the actual noise application environment. Specifically, based on analyzing the problems of information leakage in noise and "entanglement death," this research has designed a framework of high-fidelity entangled quantum teleportation channel, a quantum teleportation protocol immune to noise, and a fault-tolerant application of quantum teleportation in noise channel, so that the proposed scheme can be used in practical systems. The main work of this book is summarized as follows:

D. Li, *Quantum Teleportation Under Noise*,
https://doi.org/10.1007/978-3-031-91250-4_7

1. The framework of high-fidelity entangled quantum teleportation channel is constructed.

   Aiming at the problems of quantum decoherence and entanglement death caused by the interaction between the real physical system and the environment, a new framework of high-fidelity entangled quantum teleportation channel is designed from the aspects of entanglement death, quantum decoherence, and local common noise.

   (a) Firstly, the entanglement death problem and quantum decoherence problem in the process of quantum teleportation are analyzed, and the framework of the immune quantum teleportation channel is constructed. Since the entanglement evolution of two spatially independent atomic systems is measured by the degree of concurrency and linear entropy, we study the relationship between the initial entanglement degree of the atom and cavity field and the initial entanglement degree of the cavity field. When the initial entanglement degree of the cavity field is smaller, the duration of the first "entanglement death" is shorter. However, the two atoms and the cavity field remain entangled throughout the evolution process. In the time evolution of the entanglement of two three-level atomic systems, if it is a long-distance three-level atomic system, the rate of entanglement degradation is related to the decay rate of the atomic excited state. In the case of a close three-level atomic system, the number of single particles in the transition state determines the entanglement in the steady-state. As long as the number of single particles in the transition state is not zero, the entanglement will always exist. The entanglement before death is determined by the number of single particles in the transition state and the number of particles in the coherent superposition state of two atoms, while the entanglement after recovery is determined by the number of particles in the single-particle transition state. According to the time evolution characteristics of two spatially independent atomic systems and two three-level atomic systems, the "entanglement death" immune model is characterized by referring to the parameters related to the degree of entanglement when entanglement death occurs.
   (b) By analyzing the quantum decoherence between the atom and the cavity field in the T-C model, since the coherence of the quantum system is reflected by the off-diagonal term of the density matrix when the decoherence effect occurs, the off-diagonal term of the density matrix becomes zero, so the time scale of decoherence can measure the evolution speed of the off-diagonal term of the density matrix with time. It can also analyze the quantum decoherence by two cavity fields and two two-level atoms in the J-C model. When the dissipative system exists, the decoherence factor between the two atoms has a reduced amplitude oscillation. After a period of reduced amplitude oscillation, the amplitude decreases to zero, and a complete decoherence phenomenon will appear, that is, the dissipative system affects the decoherence of the quantum system. The evolution model of quantum decoherence entanglement is described.

(c) The influence of different noises on the entanglement evolution of entangled mixed states in the local common mode, especially when the quantum channel is affected by the local common quantum noise environment, the process of sudden death, and revival of entanglement are described, and the immune model based on density matrix and DFS joint noise is constructed.
(d) Finally, the tree graph and forest graph are encoded by the method of graph state basis under multi-degrees of freedom, and the coherence information and channel capacity are analyzed; then we can obtain the multichannel coherence information of different noise channels under the input of the encoded multi-degrees of freedom. The approximation value of the quantum capacity of the noise channel, the calculation speed, and the noise tolerance of the channel state transmission quantum information are effectively determined, and the region where quantum information can be transmitted in different noise channels is obtained.

2. The quantum information separation scheme of the Bell states and arbitrary states in different channels is proposed.

   To solve the problems of few transmitted particles and complex computation in different channels or the same channel, this book proposes to use different quantum channels to split the quantum information of arbitrary states. Firstly, the four particles' entangled state is used as the quantum channel to split the quantum information of the two-particle Bell state. After Alice sends the information to Bob and Charlie, Alice performs a GHZ state measurement on her particles, Charlie performs a single-particle measurement on his particles, and finally, Bob reconstructs the Bell state information. Secondly, in the first case, the five particles' entangled state is used to split the quantum information of any single particle and two particles. Alice performs Bell state measurement on her particles. Charlie performs single-particle measurements on his particles under the base $\{|0\rangle, |1\rangle\}$ and sends the results to Bob. Bob reconstructs the information of the original state by performing appropriate unitary operations $(I, \sigma_x, \sigma_y, \sigma_z)$. In the second case, the four-particle' cluster state and the two-particle Bell state are used to split the quantum information of any three-particle state. After Alice performs Bell state measurement on her two pairs of particles if Charlie's measurement result is $|00\rangle + |11\rangle$, Bob can restore the initial state information by performing unitary transformation $\sigma_x \otimes I$. If Charlie's measurement result is $|00\rangle - |11\rangle$, Bob can restore the initial state information by performing unitary transformation $-i\sigma_y \otimes I$. If Charlie's measurement result is $|01\rangle + |10\rangle$, Bob can restore the initial state information by performing unitary transformation $\sigma_x \otimes I$. If Charlie's measurement result is $|01\rangle - |10\rangle$, Bob can restore the initial state information by performing unitary transformation $-i\sigma_y \otimes I$. In the third case, the four-particle cluster state and GHZ state are used to split the quantum information of any three-particle state. After Alice performs the Bell state measurement on her three pairs of particles, Charlie performs a single-particle measurement on his particles under the base $\{|\pm\rangle_7\}$. Through the measurement results of Alice and Charlie, Bob performs the unitary operation (U_1, U_2, U_3, U_4) on his particles and reconstructs the information of the original state. Finally, the process of quantum information separation is physically realized in cavity quantum electrodynamics.

3. A quantum teleportation protocol for immune noise is proposed.

   At present, quantum teleportation is based on a single degree of freedom quantum state, while the real quantum physical system is a multi-degree of freedom with the characteristics of wavelength, momentum, spin angular momentum, and orbital angular momentum. In this book, a multi-degree of freedom quantum teleportation is proposed. Taking the self-spin angular momentum and orbital angular momentum of a single photon as the research object, using the state transfer characteristics of a single photon under spin angular momentum and orbital angular momentum, multi-degree of freedom teleportation in independent different noises and local common modes is achieved through the preparation of high-brightness entangled source in multi-degree of freedom, super Bell state measurement, super dense coding, and controlled identity authentication. The multi-degree of freedom characteristics of a quantum physical system are studied. The controllable quantum teleportation identity is authenticated to realize the robust quantum teleportation process. And the multi-degree of freedom quantum teleportation protocol efficiency of immune noise is analyzed and evaluated from fidelity and average fidelity.

4. Application of fault-tolerant quantum teleportation under noise.

   Aiming at the theoretical research on the model and protocol of quantum teleportation under noise, this book realizes the quantum dialogue of immune noise and quantum key distribution of continuous variables. The communication parties only need to carry out a single-photon measurement to avoid information leakage. To realize information separation under quantum key distribution with continuous variables, we randomly insert decoy states into the sequence and carry out a random translation. If the channel is secure, the key can be obtained, which will not affect the amount of information between the two parties and ensure the security of quantum key distribution.

In conclusion, the high fidelity of our proposed quantum entanglement framework addresses key challenges in quantum communication, including quantum sudden death and decoherence in transmission channels. By analyzing the evolution of quantum entanglement under local noise, we construct a noise-immune model. Finally, we implement quantum state diagram coding to enhance channel capacity. Subsequently, the quantum information separation of Bell state and arbitrary state in different channels is proposed to achieve the purpose of fast, efficient, and safe transmission of information. Then, the proposed quantum teleportation protocol for immune noise is mainly aimed at quantum teleportation in multi-degrees of freedom, quantum teleportation based on entanglement exchange, and quantum teleportation in local common mode, which solves the problems of waste of cost and low efficiency in a single degree of freedom and entanglement death in local independent mode. Finally, the fault-tolerant quantum teleportation under noise is applied to realize quantum dialogue immune to noise and quantum key distribution of continuous variables. Quantum dialogue means that both sides of the dialogue only need to carry out a single-photon measurement to avoid information leakage. The continuous-variable quantum key distribution aims at information separation by randomly inserting a decoy state and performing a random shift while inserting a decoy state and shift in the sequence is mainly in the key generation stage. If the channel is secure, the key can be obtained, which will not affect the amount of information between the two parties, thus ensuring the security of quantum key distribution.

7.2 Next Research Ideas

With the rapid development of quantum cryptography, great achievements have been made, especially in recent years. However, there are still many problems to be solved in this field. Based on this study, there are still the following directions worthy of further study.

Firstly, in the aspect of maximum entangled state degradation, we consider the method of entanglement purification to transform the non-maximum entangled state into the maximum entangled state during the protocol execution, which requires the analysis of entanglement purification to be added to the analysis of the protocol. Moreover, considering the non-maximum entangled quantum cryptographic protocol, that is, the quantum state used in the proposed protocol is no longer the maximum entangled state but directly designs and analyzes the quantum cryptographic protocol based on the non-maximum entangled state.

Secondly, quantum states are lost or changed during transmission. The loss rate of photons in optical fiber is about 0.2, and the polarization of photons is easy to change during transmission. Therefore, a quantum cryptographic protocol proved unconditionally secure in an ideal environment may not be secure in an actual environment, because quantum hackers may use factors such as quantum noise to attack quantum cryptographic protocols. In addition to quantum channels, quantum sources and detectors are not perfect in the real environment, and may also be used by quantum hackers. In addition, the most commonly used entangled states in quantum cryptographic protocols are maximum entangled states, which may degenerate into non-maximum entangled states in the process of preparation or transmission. Just because of these problems, there is a lot of follow-up work to be done.

Finally, there is no relevant solution to the problems of how to cut and flip multidimensional images under noise and achieve invisible state transfer and how to achieve efficient storage, compression, flip, and state transfer of color-saturated images, which is challenging. Of course, the quantum image is also the current research hotspot. The development of this theory also plays a vital role in promoting the development of quantum computers. This is also a very worthy research direction.

Afterword

Quantum communication has been a cutting-edge technology that has led the development of emerging industries. At present, many technical schemes of quantum communication have begun to be practical, and China has also made fruitful achievements in the field of quantum communication technology. Quantum teleportation plays an unshakable role in quantum communication. In order to promote the practicality of quantum teleportation technology, this book proposes a robust quantum teleportation protocol closer to the actual environment from the pain point of quantum teleportation. It is hoped that the contents of the book can make a contribution to enriching the theory of quantum communication, and also hope to provide new solutions in the field for more scientific researchers.

This book is supplemented and revised on the basis of my academic papers. At the same time, we need to thank Professor Liu Mingzhe for his careful guidance. Professor Liu has deep research on artificial intelligence and network security and has put forward many excellent revision suggestions for this book. Finally, I also need to thank the graduate students from the school of computer and network security of the Chengdu University of Technology: Zhou Jie, Zheng Yundan, Tan Yuqiao, Liu Xiaofang, and Yang Xiaolong for their contributions in compiling this book. This book is the result of the cooperation of all the above! I hope the publication of this book can contribute to the technology and application in the field of quantum teleportation. Due to the limited ability of the editor, there may be some inaccuracies and omissions in the book. Please criticize and correct!

D. Li, *Quantum Teleportation Under Noise*,
https://doi.org/10.1007/978-3-031-91250-4

The manufacturer's authorised representative in the EU is Springer Nature Customer Service Centre GmbH, Europaplatz 3, 69115 Heidelberg, Germany. If you have any concerns regarding our products, please contact ProductSafety@springernature.com

Printed and bound by CPI Group (UK) Ltd, Croydon, CR0 4YY
30/09/2026
02225590-0004